BREADED
FRIED FOODS

BREADED FRIED FOODS

P. KUMAR MALLIKARJUNAN

MICHAEL O. NGADI

MANJEET S. CHINNAN

CRC Press
Taylor & Francis Group
Boca Raton London New York

CRC Press is an imprint of the
Taylor & Francis Group, an **informa** business

CRC Press
Taylor & Francis Group
6000 Broken Sound Parkway NW, Suite 300
Boca Raton, FL 33487-2742

First issued in paperback 2017

ISBN 13: 978-1-138-11788-4 (pbk)
ISBN 13: 978-0-8493-1461-2 (hbk)

Library of Congress Cataloging-in-Publication Data

Mallikarjunan, Parameswarakuma.
   Breaded fried foods / Parameswarakuma Mallikarjunan, Michael O. Ngadi, and Manjeet S. Chinnan.
      p. cm.
   Includes bibliographical references and index.
   ISBN 978-0-8493-1461-2 (hard back : alk. paper)
   1. Oils and fats, Edible. 2. Deep frying. 3. Batters (Food) 4. Breading. I. Ngadi, Michael O. II. Chinnan, Manjeet S. III. Title.

TP670.M175 2009
664--dc22   2009010962

Visit the Taylor & Francis Web site at
http://www.taylorandfrancis.com

and the CRC Press Web site at
http://www.crcpress.com

Contents

Preface

Deep-fat frying remains one of the most widely used methods of preparing food. Breaded fried foods, in particular, are popular among consumers due to their unique characteristics: crispy exterior and juicy interior. Many books on deep-fat frying and fried foods have addressed this particular sector but not in a comprehensive manner, as those books covered the general concept of deep-fat frying and products such as potato chips, potato crisps, and tortilla chips. Fewer books have addressed breading and batters and the science related to coated foods in general, and some books covered only coated foods but not fried foods. We have worked on improving the quality of foods through reduction in oil uptake and enhanced quality for more than 15 years and felt there was a need to report ours as well as others' research accomplishments specific to breaded fried foods. There have been several attempts, both by researchers in academia and by the industry, to develop novel frying methods and procedures to address the challenges in developing fried foods that are attractive and at the same time perceived to be healthy. It is very important to provide an overview of those accomplishments as well.

This book addresses issues specific to breaded and battered fried foods and provides coverage of research in modeling deep-fat frying, development of novel frying systems, enhancement of the frying medium, pre- and post-frying procedures to limit oil uptake, and efforts to enhance crispness in breaded fried foods. This book will be a nice addition to the existing books on frying and fried foods, in general, and will be very useful to those in the industry involved with breaded fried foods, in particular.

We would like to acknowledge the countless number of students in our labs who worked in the area of deep-fat frying: Katarzyna Holownia, Holy Huse, Mohamed Muskat, Radha Bheemreddy, V.M. Balasubramaniam, Kulbir Pannu, Bhundit Innawong, Irina Antonova, Tameshia Ballard, Rhonda Bengtson, Lamin Kassama, Yunfeng Wang, Yunsheng Li, Lijuan Yu, Xue Jun, and Akinbode Adedeji, and our technicians who made things possible in our labs: Glenn Farrel, Sudhaharini Radhakrishnan, Vijaya Mantripragada, Richard Stinchcomb, and Aubrey Murden.

P. Kumar Mallikarjunan
Michael Ngadi
Manjeet Chinnan

The Authors

Dr. Kumar Mallikarjunan is an associate professor of food engineering in the biological systems engineering department at the Virginia Polytechnic Institute and State University. He has over 15 years of experience in the food engineering area with emphasis of his research in food process development, process modeling, nondestructive evaluation of food quality, and development of functional foods. He has more than 50 refereed publications and has received nearly $4 million for his research as principal or coprincipal investigator from agencies such as National Science Foundation (NSF), United States Department of Agriculture (USDA), United States Agency for International Development (USAID), and National Oceanic and Atmospheric Administration (NOAA). His research interests are microwave processing, deep-fat frying, far-infrared cooking, measuring quality of fried foods and frying oil, evaluation of product quality using electronic nose, Fourier transformation infrared spectroscopy, ultrasound, and extraction and encapsulation of antioxidants from agricultural byproducts and incorporation of functional ingredients in food systems.

Dr. Michael Ngadi is an associate professor in the bioresource engineering department at McGill University. He is a registered professional engineer in Canada with more than 15 years of experience in food process engineering. His research interests include cooking processes, heat and mass transfer modeling, reaction kinetics, properties of foods, and advanced emerging technologies for food processing. He is currently leading an internationally recognized research project on three major themes, namely, engineering aspects of deep-fat frying of foods (oil absorption and reabsorption, pore microstructure development), emerging food processing technologies (pulsed electric fields and intense pulsed ultraviolet light), and engineering properties of biomaterials (thermal, rheological, textural, electrical, and hyperspectral properties). His work has attracted over $1 million funding with more than 80 refereed publications. He is the recipient of several awards including the 2003 Canadian Society for Biological Engineering (CSBE) Young Engineer of the Year, the 2004 McGill's William Dawson Scholar (equivalent to the Canada Research Chair, Tier II), and the 2008 CSBE John Clark award. Dr. Ngadi teaches food engineering, biothermodynamics, material science, and linear algebra courses at McGill University. He has graduated several students at the master's and doctoral levels.

Dr. Manjeet Chinnan is a professor of food engineering at the University of Georgia. His area of expertise is in food process engineering. He is an internationally recognized authority on the processing, handling, and storage of cereal legumes, peanuts, fruits, and vegetables. He has more than 500 publications to his credit, including 130 refereed research articles, and is the recipient of more than $8 million in research grants as principal or coprincipal investigator. He is a fellow of the Institute of Food Technologist (IFT) and recipient of the 2007 Bor S. Luh International Award given by IFT. His research interests are modeling deep-fat frying operations, regenerating abused frying oil, measuring the quality of fried foods and frying oil, food process and product development optimization using mathematical and statistical modeling techniques including response surface methodology, characterizing plastic and edible films for fresh and processed foods, and microencapsulation and spray drying of phenolic and antioxidant compounds derived from plant materials.

1 Introduction

1.1 FRYING AND FRIED FOODS

Deep-fat frying is used widely around the world as a major food processing operation. In recent years, frying has become one of the fastest growing processes used in the fast food industry generating billions of dollars annually worldwide. Frying generates flavorful products that have crispy crusts, enticing aromas, and visual appeal. Owing to these unique characteristics, fried foods continue to be a major part of the prepared foods market. Consumers also consider fried foods as comfort foods. Fried chicken has been identified as one of the fastest-growing fast food menu items for the last decade and continues to grow at a rapid rate in spite of the hype about healthy, low-carb/low-fat diets (Hamaker and Panitz, 2002). This and other reports in the literature demonstrate that the consumer's vocal opinions about healthy foods and meals do not necessarily translate to their choices (Stein, 2006). Irrespective of socioeconomic background, in a study conducted in Australia with a large-scale survey of 17,531 consumers, 20% of males and 10% of females were found to consume two or more servings of fried foods per week (NSW Health Department, 2001).

Deep-fat frying technology is considered to have originated and developed around the Mediterranean area due to the influence of olive oil (Varela, Bender, and Morton, 1988). Other theories suggest that the technology developed in East Asia, mainly in a Chinese *wok* (Rossell, 2001) or in an Indian *kadhai* (Gupta, Warner, and White, 2004), and migrated to Europe. Regardless of the actual origin, today, deep-fat fried foods are found in many countries around the world.

The primary purpose of deep-fat frying is to seal the food with a crispy crust formed by immersing the food in hot oil so that all the flavors and the juices are retained. Any kind of food product can be fried uncoated or coated. An array of battered and breaded foods (cheese, fish, meat, poultry, seafood, and vegetables) represents a fast growing food category in which per capita consumption has risen from less than 5 lb in 1982 to 15 lb by 1993 (Shukla, 1993). Consumers displayed an interest toward low-fat, low-calorie products in the mid-1990s with a modest decline in fat consumption; the food industry followed suit with the introduction of over 5000 low-fat, low-calorie products. However, this decline in fat consumption was short lived. In recent years, the per capita consumption of added fats (Figure 1.1) increased from 57 lb per person during 1980 to 87.5 lb by 2004 (U.S. Census Bureau, 2007). This information is corroborated by the data on domestic shipments of edible fats and oils as reported by the Institute of Shortening and Edible Oils (Figure 1.2). Many consumers found the taste of the new low-fat and fat-free versions of foods unacceptable. In addition, America is facing an all-time high consumption of meat products. Americans now consume, as shown in Figure 1.3, an average of 73 lb of poultry and

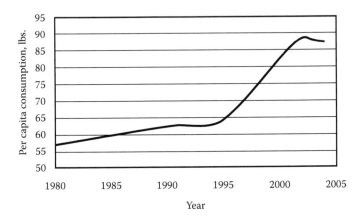

FIGURE 1.1 Per capita fat consumption for the United States. (Source: U.S. Census Bureau. 2007. Statistical abstract of the United States, Washington, D.C.)

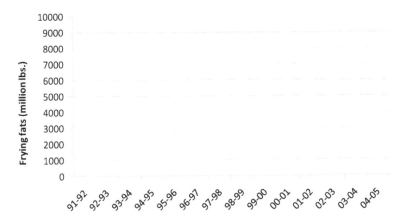

FIGURE 1.2 Per capita consumption of frying fats (ISEO, 2006. Domestic shipments of edible fats and oils as reported to the Institute of Shortening and Edible Oils. http://www. iseo.org/statistics.htm.).

17 lb of seafood per capita, which is considerably higher than they consumed in 1980 (41 lb of poultry and 12 lb of seafood) (U.S. Census Bureau, 2007). Among them, breaded fried foods like chicken and fish are particularly successful.

Harlan Sanders, well known as "Colonel Sanders," introduced breaded chicken products in the 1950s and his franchise Kentucky Fried Chicken (KFC) has nearly 10,000 restaurants in more than 40 countries worldwide. In addition to KFC, breaded fried products are very popular menu choices from many fast food restaurants and food service institutions. A whole range of deep-fat fried savory croquettes and fritters are very popular, while fruit fritters and deep-fat fried pastries are favorite desserts the world over (Table 1.1). Fast foods and foods prepared away from home are more likely to be fried and food service establishments play a major role in this. The

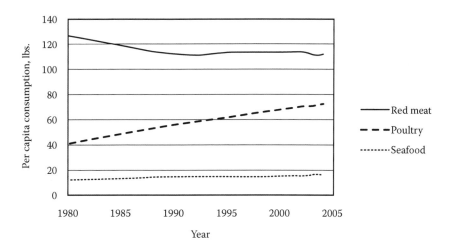

FIGURE 1.3 Per capita consumption of red meat, poultry, and seafood for the United States. (Source: U.S. Census Bureau. 2007. Statistical abstract of the United States, Washington, D.C.)

TABLE 1.1
Commonly Battered and Breaded Food Products

Seafood	Fruits
Fish sticks	Apple fritters
Fish fillets	Fried bananas
Oysters	Vegetables
Butterfly and popcorn shrimp	Onion rings
Clam strips	Bell peppers
Calamari rings	Breaded fried okra
Poultry	Eggplant
Bone-in chicken parts	Cauliflower
Marinated chicken strips (tenders)	Potatoes
Chicken patties	Mushrooms
Chicken nuggets	Zucchini
Turkey cutlets	Stuffed jalapeno peppers
Cheese sticks	Meat products
Fried ice cream	Corn dogs
Nuts	Pork fritters
	Veal cutlet patties
	Ground meat patties

number of fast food restaurants and food service establishments has been on a steady increase. The current number of such establishments is estimated to be approximately 566,000 (U.S. Census Bureau, 2007).

1.2   COATED FRIED FOODS

The increased popularity of breaded fried foods can be attributed, in part, to their textural characteristics. While consumers enjoy the crispy outer layer and the moist and juicy interior, batters and breadings also contribute to overall flavor by acting as carriers for a variety of seasonings and spices. Although breaded foods appeal to the senses, it is not their sole purpose. Breading and batters have a significant effect on the cost of the final product, reducing the cost by 20 to 30% (Sasiela, 2004).

Typically fried foods used in fast food restaurants and food service institutions are partially fried to set the breading or batter coatings and stored as frozen until frying is completed at the end by the service facilities. A schematic flow diagram of a par-frying process is shown in Figure 1.4. These partially fried breaded foods provide convenience to the restaurant or food service operators by reducing the production time required to serve the customers to within 3 or 4 min. Additionally, the partially fried breaded foods are also attractive to home users, as they need just a few minutes to cook the product without labor intensive and time-consuming preparations. The growth of the frozen food industry has also added to the benefit of delivering partially fried foods to various outlets without sacrificing quality or safety of the fried products.

Batters and breading can also be formulated to reduce oil absorption during frying, control moisture migration within the food material, prevent oxidation of the frying oil, and improve nutritive profiles (Ballard, 2003). This feature is attractive to health-conscious consumers who are often torn between enjoying fried foods and reducing fat intake. New formulations of batters and breading are being developed to carry antioxidants, micronutrients, and disease-preventing fat-soluble vitamins without diminishing the product quality.

Batter and breading technology is still considered an art but is slowly becoming a science (Corey, Gerdes, and Grodner, 1987). The functionality and ultimate success of batter and breading systems depend on the type of ingredients in the system. Typical terminologies related to breaded fried foods are listed in Table 1.2. Battered fried foods are typically exposed to rather extreme processing conditions. After application of the coating, the food product can be partially or completely cooked in frying oil at 180°C and then quickly frozen. Undesirable aesthetic problems and adverse economic impacts can occur due to partial or total loss of coating during processing, frozen storage, transportation, and handling during consumption (Mukprasirt et al., 2001).

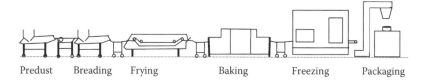

| Predust | Breading | Frying | Baking | Freezing | Packaging |

FIGURE 1.4   Typical product flow chart in a partial frying operation.

TABLE 1.2

Terminology Related to Breaded Fried Foods

Terminology	Description
Pickup	Amount of coating system on the finished product by percent
Nugget	Product having 14 to 30% pickup
Fritter	Product having 30 to 64% pickup
Croquette	Product having pickup above 64%
Par frying	A process in which the coated product has been passed through hot oil to set the batter system
Homestyle	A process in which the product has been flipped to look as if the product was done at home by hand
Tempura	A product that has no breading applied over the batter and is smooth in appearance

1.3 CHALLENGES IN COATED FRIED FOODS

The fried food industry constantly faces many new challenges and tries to address them in several fronts at once. With obesity gaining international importance, the fried food industry is trying to limit fat uptake by developing new frying technologies such as vacuum frying and infrared oilless frying. In addition, new breading and batter formulations are being examined to limit fat uptake and at the same time enhance crispness in fried foods. The other challenge the industry has is maintaining and enhancing the shelf life of fried foods under the food warmer (or heat lamp) because any extension of quality under a heat lamp will increase the profit potential for many food service outlets. Research is also focusing on the acrylamide in fried foods even though recent research has suggested the amount of acrylamide in fried foods may not be considered as a health risk. Deli counters also face challenges related to reheating food products, especially lipid oxidation and warmed over flavor in meat products. In manufacturing partially fried foods, the industry is investigating ways to improve the yield in terms of breading pickup for economic benefits.

1.4 REFERENCES

Ballard, T. 2003. Application of edible coatings in maintaining crispness of breaded fried foods. Masters Thesis. Virginia Polytechnic Institute and State University.

Corey, M.L., Gerdes, D.L., and Grodner, R.M. 1987. Influence of frozen storage and phosphate predips on coating adhesion in breaded fish portions. *J. Food Sci.* 52(2): 297–299.

Gupta, M.K., Warner, K., and White, P.J. 2004. *Frying technology and practices*, Urbana, IL: AOCS Press.

Hamaker, S.S. and Panitz, B. 2002 In vogue: What's hot in the restaurant industry, *Restaurant USA*, May issue. http://www.restaurant.org/rusa/magArticle.cfm?ArticleID=768 accessed September 14, 2007.

ISEO. 2006. Domestic shipments of edible fats and oils as reported to the Institute of Shortening and Edible Oils. http://www.iseo.org/statistics.htm, accessed September 14, 2007.

Mukprasirt, A., Herald, T.J., Boyle, D.L., and Boyle, E.A.E. 2001. Physicochemical and microbiological properties of selected rice flour-based batters for fried chicken drumsticks. *Poultry Sci.* 80: 988–996.

NSW Health Department, Sydney. 2001. http://www.health.nsw.gov.au/PublicHealth/surveys/hsa/9798/nut/nsw_nut_nsw_nut_n8_ses.htm. Accessed September 14, 2007.

Rossell, J.B. 2001. Factors affecting the quality of frying oils and fats. In *Frying: Improving quality*. Rossell, J.B., Ed., Cambridge, U.K.: Woodhead Publishing Limited, pp. 115–164.

Sasiela, R.J. 2004. Technology of coating and frying food products. In *Frying technology and practices,* Gupta, M.K., Warner, K., and White, P.J., Eds., Urbana, IL: AOCS Press.

Shukla, T.P. 1993. Batters and breadings for traditional and microwavable foods. *Cereal Foods World.* 38(9): 701–702.

Stein, R.L. 2006. Healthful fast foods are not part of healthful revenue, *J. Am. Dietetic Assoc.* 106(3): 344–345.

U.S. Census Bureau. 2007. *Statistical abstract of the United States*, Washington, D.C. pp. 133, 780.

Varela, G., Bender, A.E., and Morton, I.D. 1988. *Frying of food: Principles, changes, new approaches*. Chichester, U.K.: Ellis Horwood Ltd.

2 Principles of Deep-Fat Frying

2.1 FRYING AS A UNIT OPERATION

Deep-fat frying is used widely around the world as a major food processing operation. In recent years, frying has become one of the fastest growing processes used in the fast food industry generating billions of dollars annually worldwide. Despite its popularity, frying is still very much an art and the theoretical aspects of the process are highly complex and continue to present difficulties to scientists. There are a series of different phenomena occurring simultaneously during a deep-fat frying process. In particular, there is simultaneous heat, moisture, and fat transfer taking place between the product and the heating medium (frying oil). The formation of a crust layer on the outer surface of the product is another characteristic event that takes place. To complicate issues even further, the composition of the oil and the properties of the products are steadily changing throughout the process. It is important to have a good understanding of these activities during the frying process in order to optimize and control the process. Optimization allows for simplification of the control documentation aspects of the process. This should translate to several benefits including increased frying oil life, decreased oil absorption by the product, decreased product rejection rates through tightened process control specifications, energy conservation, and reduced operating costs (Blumenthal, 1991). Deep-fat frying operation involves immersing a food product into a hot-oil bath for a given period of time. Oil temperatures are typically in the range of 175 to 205°C (350 to 400°F), resulting in moisture loss due to cooking of the food product and evaporation. Fat uptake by the food product occurs simultaneously with moisture loss. For a breaded product, there is the bread coating at the surface and an inner substrate product core which changes the nature of heat, moisture, and fat transfers. Rapid heat transfer from the frying oil quickly sets the structure of the coating, allowing only limited moisture loss and fat uptake. A crust forms at the surface during frying, especially within the breading region of the food where there is rapid evaporation of water. The hardening of the surface locks moisture inside the product, resulting in a relatively moist interior and dry exterior. Thus, the coating is a barrier to mass transfer during frying. The bread crust is one of the most important characteristics that give fried foods their unique crunchy texture.

The oils used in deep-fat frying provide flavor, increase the caloric content of the food, and contribute nutritional and physiological elements, including fat-soluble vitamins, essential fatty acids, prostaglandin precursors, caloric energy, and satiation (Perkins and Erikson, 1996). The frying oil is subjected to thermal and oxidative reactions as it is heated at high temperature in open air (White, 1991). These degenerative reactions affect the viscosity of the frying medium over time. Consumption of high amounts of oil and its frying by-products is a cause of public concern due to links with cardiovascular disease, obesity, colon cancer, and other disorders (Pinthus et al., 1995). The National Heart Lung and Blood Institute (NHLBI) recommends that the caloric intake of fats and oils should not exceed 30% of daily energy needs (Perkins and Erikson, 1996). Accordingly, there is increasing interest in reducing fat absorption and fat content of fried foods. Application of coatings and breadings can be seen as an ingenuous strategy in reducing fat absorption during frying. Information available from several studies shows that coatings are effective in modifying moisture loss and fat absorption in frying products. However, careful selection of suitable coatings that conform to appropriate food regulations and meet the desired quality requirements is necessary.

2.2   HEAT TRANSFER

Heat is transferred from oil to a frying food product, resulting in moisture evaporation and changes in the food product. Since most food products have high moisture contents, moisture and its evaporation play a variety of roles in defining heat transfer during deep-fat frying. Moisture carries off thermal energy from the hot frying oil surrounding the frying food. The bubbling water at the surface of the product influences heat transfer coefficient in ways that have not been fully elucidated. As water escapes from the inner portion of the product and comes into contact with the hot oil, bubbles form and move vigorously throughout the oil, thereby causing turbulence (Innawong, 2001). In general, turbulent conditions promote more rapid heat transfer. The amount of water vapor bubbles decreases with increased frying time due to the decreased amount of remaining moisture within the product. Evaporation of moisture from the food product also has implications on oil absorption.

The removal of energy from the surface of foods prevents charring and burning that would have been caused by the high frying temperature and excessive dehydration. Thus, the temperature of most of the interior of a frying food is normally at approximately 100°C (the temperature of phase change of water) for the majority of the frying time. From a heat transfer point of view, deep-fat frying can be divided into two major regimes: namely, non-boiling, when the temperature of the product is below the boiling temperature of water, and boiling. Characterization of these regimes is vital in determining required frying time for a given product. Farkas (1994) expanded these two frying regimes into four stages as follows:

1. Initial heating stage, lasting for only a few seconds when the surface of a food heats to a temperature equivalent to the elevated boiling point of liquid before vaporization initiates.
2. Surface boiling stage, when vaporization begins at the surface and a crust begins to form.
3. Falling rate stage, when the internal core temperature rises and more internal moisture migrates outward and out from the food. Most thermophysical changes such as starch gelatinization, protein denaturation, and

cooking occur in the inner core of the food at this stage. In addition, a crust is formed at the surface with increasing thickness at longer frying time.

4. Bubble end-point, which is the final stage after a considerably long period of frying. The rate of moisture removed decreases and no more bubbles are seen escaping from the surface of the product.

Stages 2 to 4 can be combined to represent the boiling regime. Most foods are sufficiently cooked by Stage 3.

There are two basic modes of heat transfer involved in the process of deep-fat frying. These are convection and conduction modes of heat transfer. Heat is transferred from the oil to the surface of a product by convection, whereas conductive heat transfer takes place within the food material. There are two commonly accepted classical theories on the mechanism of conductive heat transfer. The first elaborates that when molecules of a solid material attain additional thermal energy, they become more energetic and vibrate with increased amplitude of vibration while confined in their lattice. These vibrations are transmitted from one molecule to another without any molecular motion. Thus, heat is conducted from a higher temperature area to a lower temperature area. The second theory states that conduction occurs at a molecular level due to the drift of free electrons. These free electrons are prevalent in metals, and they carry thermal and electrical energy. For this reason, good conductors of electricity such as copper are also good thermal conductors. Obviously, the first theory is more relevant for conductive heating of food products since foods do not typically possess free electrons as metals do. Simple heat conduction in a product can be expressed as follows:

$$\rho c_p \frac{\partial T}{\partial t} = \nabla.\left(k\nabla T\right) \tag{2.1}$$

where T is temperature (°C), ρ is product density (kg/m^3), c_p is heat capacity (J/kg°C), k is thermal conductivity, and ∇ is the nabla gradient. The nabla gradient can be expanded for different coordinates. For instance, the nabla gradient in a three-dimensional Cartesian coordinate can be expressed as follows:

$$\nabla T = \left(\frac{\partial T}{\partial x}, \frac{\partial T}{\partial y}, \frac{\partial T}{\partial z}\right) \tag{2.2}$$

Conduction heat transfer is influenced by the thermal properties of the food such as thermal conductivity, thermal diffusivity, specific heat, and density. Convective heat transfer occurs when a moving fluid comes into contact with a solid at a different temperature. In deep-fat frying, convection takes place as the circulating oil contacts the frying food. The temperature gradient between the food product and the frying medium is the driving force for convective heat transfer as described by Newton's law of cooling [Equation (2.3)].

$$q'' = h(T_\infty - T_s) \tag{2.3}$$

where q'' is convective heat flux (W/m^2), h is the convective heat transfer coefficient (°C), T_∞ is oil temperature, and T_s is surface temperature.

The convective heat transfer coefficient is the major factor influencing heat transfer rate to the frying product. It relates mainly to the property of a transport boundary layer developed between the frying medium and the product. In general, convective heat transfer can be in either forced or free mode. Forced convective heat transfer occurs when the heating fluid is forced to flow around the heated solid. For example, a pump can be used to circulate frying oil through fryer and food. In free (or natural) convection, the bulk fluid motion is due to differences in density or buoyancy effects. For example, a food may be heated by stagnant frying oil. In reality, there may be a mixture of the free and forced convective modes of heat transfer during frying. A complete description of heat transfer during frying will include both the conductive and convective heat transfer modes. Therefore, all thermal properties of the food material, namely, specific heat, thermal conductivity, and density as well as convective heat transfer coefficient between the product and the oil, are important parameters that define the rate of heat transfer. These thermo-physical properties of the food material continually change due to moisture depletion and temperature change (Costa et al., 1999; Hallstrom et al., 1988). Data on thermal properties of most foods can be obtained from the literature (Mohsenin, 1980; Rao et al., 2005).

Accurate determination of thermal properties of foods is critical for reliable mathematical modeling. Thermal properties of meats from room temperature to frozen temperature range are readily available (Sanz et al., 1987). Only limited data are available for frying conditions. Ngadi and Ikediala (1998) estimated the specific heat of chicken drum muscle at different moisture contents (up to 80% w.b.), fat contents (15 to 18% d.b.), and temperatures (25 to 125°C) using a differential scanning calorimeter. The values of specific heat ranged from 1.56 to 4.08 kJ/(kg.K). Miller et al. (1994) estimated the convective heat transfer coefficients during frying using a lumped capacitance method. At 188°C oil temperature, the heat transfer coefficient for coconut and canola oil were reported to be 288 and 282 W/(m^2.K), respectively. Ngadi and Correia (1994b) estimated heat transfer coefficients for chicken drumstick-shaped bodies during deep-fat frying and the reported values ranged from 67 to 163 W/(m^2.K).

Information on thermo-physical properties of batter coating is scarce. It is difficult to accurately quantify heat transfer coefficients around a food product during a deep-fat frying process. Data on the convective surface heat transfer coefficients (h-value) obtained during food frying through the use of model metal transducers, food, and model foods have been reported in the literature (Ngadi and Ikediala, 2005; Sahin et al., 1999; Costa et al., 1999; Pannu and Chinnan, 1999; Dincer, 1996; Miller et al., 1994). The general approach of using a metal transducer is to measure the center temperature (or the temperature at a specified location) while the transducer heats or cools in oil. Thus, the heat transfer coefficient may be estimated from the time-temperature data. Breaded foods may show somewhat higher heat transfer coefficients due to their relatively uneven surfaces. However, this may be difficult to verify. Another approach that has been used to estimate the heat transfer coefficient is to systematically adjust estimated values until the error between experimental and calculated time-temperature data for a frying product is minimized. Heat transfer coefficients can also be estimated from experimental moisture evaporation rates by assuming that heat transferred to the frying product is used in evaporating moisture.

$$q = \Delta H_{vap} \frac{dm}{dt} \tag{2.4}$$

where q is heat transferred (W), ΔH_{vap} is heat of vaporization (J/kg), m is moisture, and t is frying time (s). Thus:

$$h = \frac{dm}{dt} \frac{\Delta H_{vap}}{A(T_o - T_s)} \tag{2.5}$$

These methods assume that an appropriate heat transfer model is used and that the experimental data are reliable and consistent. In practice, heat transfer coefficients can be based on product shape and size in addition to fluid characteristics.

Data in literature suggest that frying temperature, oil viscosity, heat capacity, and surface tension are the primary physical properties that may influence the heat transfer coefficient and product heating rate (Blumenthal, 1991; Miller et al., 1994; Sosa-Morales et al., 2006; Yldz et al., 2007). Blumenthal (1991) outlined the surfactant theory and suggested that the heat transfer rate from oil to food can change as the oil quality degrades because of prolonged use. It was noted that heat is transferred during food frying from a non-aqueous medium (frying oil) into a mostly aqueous medium (food product). There is an oil–water interface that does not exist when metal transducers are used to ascertain the h-value. Therefore, metal transducers may not register changes in h-values due to the buildup of surfactants. This is supported by results presented by Tseng et al. (1996) when they showed that the h-value is not affected until the oil discard point is reached. Miller et al. (1994) used h-value data to study the influence of temperature and oil type on oil use (abuse) and found that the major effect was due to viscosity changes. However, the differences in h-values were small. Budzaki and Seruga (2005) used moisture rate data to estimate heat transfer coefficients during deep-fat frying of dough. The authors reported that convective heat transfer coefficient followed similar trends as moisture rate. A maximum heat transfer coefficient was attained early during frying when moisture rate was high. The heat transfer coefficient subsequently decreased to minimum values when moisture loss became constant. Despite these difficulties, average heat transfer coefficients obtained using different techniques appear to be in the same order of magnitude of 100 to 400 W/m²°C. However, heat transfer coefficient values obtained with metal transducers may be used with caution as long as the food product is fried in good quality oil.

2.3   MOISTURE TRANSFER

In general, mass transfer during frying encompasses moisture and oil transfer. Moisture evaporation occurs initially at the product surface and then later at the interface between the dry layer at the surface and the wet core of the food (Levine, 1990). A diffusion gradient between the dry surface and the wet core of the food as well as the pressure gradient created by the evaporation of the inner moisture are responsible for moisture loss during frying (Gamble and Rice, 1987). Evaporated moisture and steam find "selective weaknesses" in the structure of the product to escape as bubbles while frying proceeds. The amount of water vapor bubbles expelled from the food decreases over time due to the decreasing moisture content in the food (Singh, 1995). Thus, the rate of moisture loss is typically rapid within the first 60 s of frying, but it reduces dramatically afterward and approaches a constant value toward the end of frying. A fundamental approach to describing moisture transfer during deep-fat frying is to assume that the process is diffusion controlled. Hence, Fick's second law of diffusion is the governing equation for moisture transfer (Ngadi et al., 1997).

$$\frac{\delta(\rho_s m)}{\delta t} = \nabla(D\rho_s \nabla m) \tag{2.6}$$

where m is moisture content on a dry basis (kg/kg), ρ_s is the density of dry solids (kg/m^3), t is time (s), and D is effective moisture diffusivity (m^2/s).

If moisture diffusivity and density are constant, Equation (2.6) will be reduced to:

$$\frac{\delta M}{\delta t} = D\nabla^2 M \tag{2.7}$$

Defining appropriate boundary conditions and solutions for the moisture transfer equation may prove difficult especially for a composite coated product such as breaded food. In terms of average product moisture content during frying, several authors have established that the instantaneous rate of moisture loss at any time during frying is proportional to the moisture content at that instant and it increases with increasing frying oil temperature (Costa et al., 1999; Krokida, Oreopoulou, and Maroulis, 2000; Krokida et al., 2000, 2001; Budzaki and Seruga, 2004; Ngadi et al., 2006; Taiwo et al., 2007). This has allowed the use of first-order reaction kinetics to describe moisture loss during frying. Several other studies have simply correlated moisture loss with frying time and it has been suggested that moisture loss is proportional to the square root of frying time (Mittlelman et al., 1982; Ashkenazi et al., 1987; Gamble et al., 1987; Rice and Gamble, 1989). The batter/breading (coating) systems provide protective coating that minimizes moisture loss. The core portion of breaded foods normally experiences more gradual decrease in moisture content. Rapid drying is critical for ensuring desirable texture of the final product. However, it is undesirable to have excessive moisture loss as it may result in greater absorption of oil by the product or poor texture. Bread coatings serve to maintain higher moisture content in the final product normally, which also results in lower final fat content. Performance of bread covering during frying is strongly affected by additives and ingredients in the batter.

2.4   FAT TRANSFER

Oil absorption into a product, during frying, is influenced by oil temperature, frying time and surface moisture content, product surface area, and pressure (Innawong, 2001). Gamble and Rice (1987, 1988) and Rice and Gamble (1989) reported that oil absorption and moisture content of fried products were directly related. Several researchers have reported that the oil content in the product is independent of the frying temperature but closely related to the water content (Krokida et al., 2000; Gamble et al., 1987; Pinthus and Saguy, 1994; Rice and Gamble, 1988).

Moisture loss creates cavities or pores as well as passageways in a food. These cavities are known as capillary pores and through them oil penetrates fried products during frying. Thus, predicting water loss is critical for modeling and controlling deep-fat frying operations. Increased frying temperature generally decreases oil uptake because of the reduction in the overall time in the fryer (Gamble et al., 1987). Average oil uptake, as with moisture loss, has been expressed as a function of the square root of frying time. The total yield of the product decreased with frying time because the loss of moisture was more rapid than oil uptake. The distribution and the absorption of oil depend on several factors including the pre-drying treatment, frying time, surface area, and the thickness of the products. Guillaumin (1988) reported a linear relationship between thickness of potato chips and the amount of oil absorbed. Gamble et al. (1987) and Gamble and Rice (1988) reported that when frying 1.04- to 2.11-mm thick potato slices, the oil uptake was related to the surface area of the potato slices. The final oil content decreased linearly with increased thickness. Moreira et al. (1991) also reported that the distribution of the oil in tortilla chips was not uniform and that most of the oil easily concentrated around the edges and the puffed area of the chips.

Capillary displacement plays a central role in defining fat absorption during frying. Saguy and Pinthus (1995) proposed mechanisms of moisture loss and oil absorption during deep-fat frying as the following: (1) high temperature creates "explosive" boiling of the water contained in the fried material; (2) this bursts cell walls and forms capillary holes and voids; (3) oil is absorbed into those holes and voids; and (4) oil uptake is increased by a reduction of internal pressure due to water loss and by subsequent cooling, which creates a "vacuum effect." The authors further suggested that the oil that enters the voids left by the moisture loss could hold the capillaries open by keeping the structure from shrinking or collapsing. This hypothesis may be supported by the Attenuated Total Reflection Fourier Transform Infrared (ATR-FTIR) scan of a fried product that had been dried, shown in Figure 2.1. Fat could be seen occupying parts of the pores that presumably had been vacated by moisture. Thus, moisture loss could be also affected by the amount of oil uptake.

Most of the work done on deep-fat frying in relation to oil transfer has been limited to non-fatty foods such as potato products and tortilla chips. Few researchers

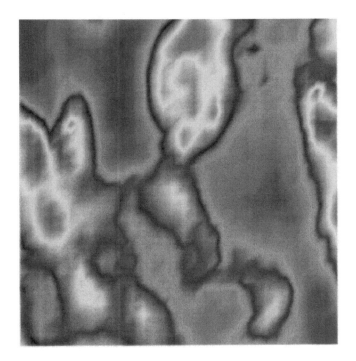

FIGURE 2.1   Fat profile image from FTIR.

have studied the mechanism of fat transport in initially fatty products such as meat and chicken products. Ateba and Mittal (1994) developed a modeling simulation of simultaneous fat, heat, and moisture transfer during deep-fat frying of beef meatballs having an initial fat content. They proposed that fat transfer could be divided into two periods including fat adsorption and fat desorption. During the fat absorption period, oil diffuses into the product, whereas the fat desorption period is marked by the migration of fat from the product to the surroundings. Fat is absorbed to the product surface into the space vacated by evaporated moisture. When the moisture is evaporated from the surface, void spaces are left behind in the product. Fat is then absorbed and fills those void spaces. This phenomenon of fat replacing escaping moisture was also suggested by Guillaumin (1988). Foods lacking an initial fat content do not experience the fat desorption period. Pressure difference was thought to be the most probable reason for fat desorption. The driving force is taken to be tension gradient in the capillaries. This force is caused by the expansion of fat and shrinkage of the capillaries in the product probably due to collagen shrinkage and denaturation of meat proteins. Mallikarjunan et al. (1995) also reported similar absorption and desorption periods occurring at the surface of deep-fat fried chicken nuggets.

2.5   MOISTURE AND FAT TRANSFER IN COATED PRODUCTS

Coating modifies moisture and fat transfer characteristics of foods during deep-fat frying as shown in Figure 2.2. For uncoated products, the crust layer formed at longer

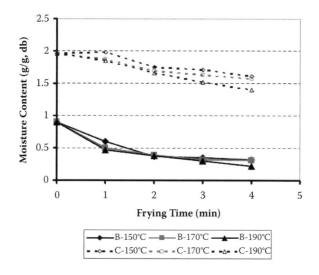

FIGURE 2.2 Moisture profile in the breading (B) and core (C) regions of chicken nuggets during oven baking at different temperatures. *Source*: Ngadi, M., Dirani, K., and Oluka, S. (2006). Mass transfer characteristics of chicken nuggets. *Intl. J. Food Eng.*, 2(3): 8, 1–16..

frying time reduces moisture loss. However, since the coating sets and its equilibrium fat content is attained quickly at the high frying temperatures, there is more reduction of moisture loss and fat absorption in breaded and coated products. Edible coatings prepared from ingredients such as polysaccharides, protein, lipids, or their combinations can be used as part of batters and breading to improve batter and coating performance in terms of reducing fat absorption and moisture loss (Kester and Fennema, 1986; Debeaufort et al., 1998; Dziezak, 1991; Sanderson, 1981).

By suitable selection of coating material, it is possible to control moisture and fat transfer between the frying medium and the food (Mallikarjunan et al., 1997). Ngadi et al. (2006) studied the pattern of moisture loss and fat absorption in breaded chicken nuggets. These patterns of moisture loss and fat absorption are typical for most other breaded products. Breading moisture loss was influenced by frying time and temperature. The breading moisture decreased at all frying temperatures following a pattern that was peculiar to most drying behavior. For chicken nuggets, the drying curve pattern exhibited an initial constant drying rate period lasting for about 1 min followed by a dropping rate period between 1 and 2 min, then a second constant rate period from 2 to 4 min of frying. The second constant rate period showed a lower drying rate than the initial constant rate period. This moisture loss pattern was directly associated with the composite structure of chicken nuggets. The two initial drying stages of constant and then dropping rates are due to rapid moisture loss from the breading portion. The last stage, which was referred to as a second constant-rate drying phase, was due to the onset of moisture loss from the core portion of the chicken nuggets. This moisture, transferred from the core, had to pass across the breading layer. Thus, the breading moisture content was affected by the moisture lost from the core at the later stages of frying.

Moisture content in the core portion of breaded products also decreases with increasing frying time and temperature. For frying chicken nuggets at temperatures between 150 and 190°C, Ngadi et al. (2006) reported that moisture content of the core portion did not change appreciably within 1 min of frying, but only decreased slightly at higher frying times. In general, moisture content in the core part of the chicken nuggets decreased with frying time at a lower rate than the moisture content of the breading. This low rate of moisture loss is attributed to the protective coating provided by the batter and breading layer surrounding the core. Moisture loss from the core region tends to follow a linear trend. This indicates a constant rate of drying, similar to the early constant drying rate observed in the breading layer. However, the drying rate observed in the core portion was lower than that observed in the breading portion at the early stages of frying. The drying behavior of the core was expected to exhibit a dropping rate stage upon further frying time. The results obtained for moisture loss from the core portions of chicken nuggets are comparable with moisture loss observed by Indira et al. (1999) for the stuffing portion of fried samosa samples. The stuffing moisture content remained at approximately 63% (wb) throughout frying regardless of frying temperature. The unchanged moisture content of the stuffing portion of the samosa was due to the protective and sealing function of the casing, which is comparable with the role of the breading/batter portion of chicken nuggets in retarding moisture loss from the core as emphasized by Suderman (1983) and Davis (1983).

Moisture loss and fat absorption patterns in breaded foods can be modified further by incorporating film forming hydrocolloids as part of batters and breading (Kester and Fennema, 1986; Debeaufort et al., 1998; Dziezak, 1991; Sanderson, 1981). Methylcellulose (MC) and hydroxymethylcellulose (HPMC) possess good film forming characteristics. Their films are generally odorless and tasteless, flexible and are of moderate strength, transparent, resistant to oils and fats, water-soluble, moderate moisture, and oxygen transmission (Krochta and de Mulder-Johnston, 1997; Nisperos-Carriedo, 1994). MC and HPMC have the ability to form thermally induced gelatinous coating. Thus, they have been used to retard oil absorption in deep-fat frying food products (Baker et al., 1994; Balasubramaniam et al., 1997; Dziezak, 1994; Kester and Fennema, 1986; Mallikarjunan et al., 1997; Nisperos-Carriedo, 1994; Sanderson, 1981). Balasubramaniam et al. (1995, 1997) demonstrated that there is the potential of using edible films for moisture retention and reduction in fat absorption during frying of poultry products. Compared to "control" uncoated samples, the coating of edible films on chicken balls (nuggets) reduced the fat absorption in the surface layer up to 17.9% and in the core up to 33.7%. Mallikarjunan et al. (1997) studied the moisture retention and fat reduction capabilities of different edible film coatings during deep-fat frying of starchy products. It was shown that moisture reduction was 22 and 31% for MC and HPMC, whereas the reduction in fat uptake was 83.6 and 61.4% for MC and HPMC, respectively.

The ability of these films and coatings to limit moisture transfer may be the key to the production of crispier breaded fried products. Furthermore, edible films and coatings, by acting as barriers to control the transfer of moisture, oxygen, carbon dioxide, lipids, and flavor compounds, can prevent quality deterioration and increase the shelf life of food products (Mate and Krochta, 1996). As a result, fried foods that are coated with edible films are more moist food products and the films aid in extending the fry life of the oil. Gas and water vapor barrier properties of an edible film and coating vary greatly with composition and presence of bubbles and pinholes of the films (Park and Chinnan, 1995).

The chemical and physical structure of edible film coatings is what makes them such effective barriers against oil and moisture. The use of MC and HPMC for their oil and moisture barrier properties has been more widely investigated and reported than the use of any of the other hydrocolloids (Loewe, 1990). MC and HPMC, which are cellulose-based hydrocolloids, exhibit reversible thermal gelation, which causes batters to "set" temporarily during frying. As a result, they reduce batter "blow off" and pillowing and decrease residual debris in cooking oils (Mukprasirt et al., 2000). The ability to reduce residual debris in the frying oil is an added advantage of using HPMC and MC in frying. A reduction in batter debris aids in preserving the quality of the fry oil. The removal of debris from the oil is essential because the debris imparts undesirable flavor and color compounds into the oil, causing it to darken.

Cellulose gums regulate batter viscosity and aid in reducing oil uptake and controlling moisture retention within the food material. In a study conducted by Mallikarjunan et al. (1997), it was shown that mashed potato balls coated with MC, HPMC, and corn zein (CZ), in comparison to the control, had moisture loss reductions of 31.1, 21.9, and 14.5% for MC, HPMC, and CZ, respectively. Accordingly, reductions in fat uptake were 83.6, 61.4, and 59% for MC, HPMC, and CZ, respectively. It was also found that among the films tested, MC exhibited the best barrier properties to provide moisture retention and reduction in fat uptake during deep-fat frying. The better moisture barrier performance of MC coatings compared to HPMC coatings was attributed to MC being less hydrophilic than HPMC. In a Dow Chemical Co. study (1991), batters formulated with HPMC absorbed 26% less oil than the control after a 2-min fry cycle. Furthermore, addition of HPMC as a pre-hydrated gum solution to the batter resulted in an even greater oil reduction; up to 50% more in some applications.

Whey protein, a byproduct of the cheese industry, has excellent nutritional and functional properties and the potential to be used for edible films (Mate and Krochta, 1996). When wheat gluten is added to a batter mix, its film-forming properties reduce moisture loss and produce crisp, appetizing surfaces. In a study of the effect of edible coatings on deep-fat fried cereal products, soy protein isolate, whey protein isolate, and MC were found to be the most effective moisture and fat barriers (Albert and Mittal, 2002). Additionally, pre-dusting food with wheat-based films significantly improved adhesion and enhanced the appearance (Magnuson, 1985). Whey protein contains lactose, which is a reducing sugar involved in browning reactions that impart more color to the food material (Loewe, 1990). The fried food industry will continue to seek new and improved coating materials that result in reduction of fat absorption and calories as well as infusion of health and flavor-related attributes.

2.6   CRUST FORMATION

The formation of a golden brown, crispy layer on the outer surface of the product is perhaps the most recognizable characteristic of fried foods. This layer, known as the crust, is formed within minutes after the product comes into contact with the oil through both chemical and structural changes in the product. The golden color of the crust can be attributed to Maillard reactions involving chemical changes in the sugar compounds on the product surface. Low water content in combination with high temperatures causes Maillard reactions to occur (Skjoeldebrand and Olsson, 1980). The crust is a dry layer that acts as a barrier between the inner portion of the food material and the surrounding oil. Due to its dry nature and inability to efficiently conduct heat, the crust becomes heat transfer limiting. Not only does the development of the crust influence heat and mass transfer but it influences oil uptake as well. Oil uptake during deep-fat frying is mostly localized in the crust. Oil tends to concentrate near edges, corners, and broken "slots" (Pinthus and Saguy, 1994). As the crust layer begins to thicken as a result of increased frying time, it no longer permits oil to be passed through. Accurate measurement of crust thickness in fried foods can be challenging. Pinthus et al. (1995) evaluated the thickness and internal structure of fried potato slices by sectioning fried products and separating the crust and core portions. Thickness, porosity, and gel strength of the crust section were measured and related to frying parameters. The authors reported that crust thicknesses of the products were in the range of 327 to 650 μm, increasing with increasing frying time as expected. The porosity of the crust region decreased linearly with frying time. Normally, increase in porosity of the fried part of the product is expected to increase oil uptake. The decreasing porosity observed by the authors was due to the intrusion of oil into the pores and voids created by evaporated water. Crust porosity and fat uptake were directly related to the initial deformability modulus of the unfried product. Crust porosity decreased, whereas fat uptake increased for lower deformability modulus of the unfried product. Thus, it may be possible to adjust a batter coating system in order to control porosity and fat absorption during frying. Ateba and Mittal (1994) used simulated temperature profiles from mathematical modeling to estimate crust thickness of approximately 2.5 mm during deep-fat frying of 47-mm diameter meatballs. It was assumed that crust formed at the regions where temperatures were more than 100°C. The crust thickness reported by Ateba and Mittal (1994) for the meatball was thicker than that reported for potato slices by Pinthus et al. (1995). Crust development in fried foods has vital influence on fat absorption and texture. The characteristics of the crust may depend on the type and nature of the product being fried.

2.7   MATHEMATICAL MODELING

To better control the deep-fat frying process, optimize heat and moisture transfer in breaded products, and effectively reduce oil uptake in breaded products, a better understanding of heat and mass transfer parameters in addition to breading characteristics is necessary. Lane et al. (1980) evaluated the effect of internal temperature of breaded deep-fat fried chicken thighs on the sensory perception of "doneness."

They found that a minimum of 14.5 min of deep-fat frying at 163°C was necessary to reach an internal temperature of 93°C, at which a trained panel judged the product as "done." Useful information about minimum temperatures at different frying conditions can be obtained with appropriate mathematical models, thus optimizing processing conditions and reducing product development time.

There have been several attempts to simulate the frying process based on mathematical principles. Mathematical models may be product specific due to product properties including differences in the product's geometry and chemical and physical makeup. To develop an appropriate model, it is vital to recognize that deep-fat frying is a complicated dehydration process involving transfer of heat and different mass species in ways that are still being elucidated. In this regard, assumptions are made, which simplifies the resulting mathematical formulations. The applicability of the mathematical model depends on the validity of the assumptions and how closely related they represent actual frying situations. Available mathematical models can be divided into two general types; namely, single and composite regions models. For the latter case, the model assumes a moving boundary at the crust/coating interface unlike the first in which the frying product is assumed to be one single region. Different authors have used these types of models with different levels of complexities. A region may be treated as a homogenous single-phase medium or a multiphase porous medium comprising solid, liquid, and gas matrixes. Early attempts to model deep-fat frying used simplified single layer models and focused on describing the individual transfer process (i.e., heat, moisture, and fat transfer) during the frying process (Gamble et al., 1987; Rice and Gamble, 1989). More recent models of deep-fat frying typically consider either multilayers or composite structure for the frying product (Farkas et al., 1996a,b; Costa and Oliveira, 1999).

2.7.1 Single Layer Models

A very simplistic approach to modeling deep-fat frying is to assume negligible external resistances to heat, moisture, and fat transfer phenomena in a product during frying. This approach may be sufficient for thin, crustless products (soy protein, tofu, potato slices, extruded puffed snacks, etc.) with no appreciable internal temperature gradient. For this scenario, convective heat transfer can be equated to change in the heat content of the product and temperature profile can be estimated following the lumped parameter approach (if $Bi < 0.1$):

$$T_r = \frac{T(t) - T_\infty}{T_o - T_\infty} = e^{-F_o t} \tag{2.8}$$

where

$$F_o = \frac{hA}{mc_p} \tag{2.9}$$

The assumption of negligible external resistance is rarely valid for heat transfer during deep-fat frying except for products small enough to satisfy the $Bi < 0.1$

condition. Further, the surface temperature of the product changes considerably such that the heat transfer coefficient may not be high enough to justify the assumption of a negligible external resistance. A more realistic approach for a single layer material would be to formulate a heat conduction equation based on a representative coordinate. This equation is then solved with the appropriate boundary and initial conditions. For instance, for a homogenous and isotropic material, the 1-D heat conduction equation may be written as

$$\frac{\partial T}{\partial t} = \alpha \frac{\partial^2 T}{\partial x^2} \tag{2.10}$$

where α is thermal diffusivity (m²/s).

Initial and boundary conditions are required to solve the heat conduction equation. Assuming uniform initial temperature, and convective heat transfer at the surface, the following boundary condition equations can be written:

$$T(x,o) = T_o \tag{2.11}$$

$$k \frac{\partial T}{\partial x} = h(T_\infty - T_s) \tag{2.12}$$

$$k \frac{\partial T}{\partial x} - D_e \rho_d L \frac{\partial m}{\partial x}\bigg|_{x=X} = h(T_\infty - T_s) \tag{2.13}$$

Equations (2.10) through Equation (2.13) are solved to obtain spatial temperature profiles in the material.

Fick's second law of diffusion can be used to describe mass transfer in a 1-D homogenous and isotropic product.

$$\frac{\delta m}{\delta t} = D \frac{\delta^2 m}{\delta x^2} \tag{2.14}$$

Assuming a negligible external resistance, uniform initial moisture content, and equilibrium moisture content, the solution for the Equation (2.14) can be obtained from Crank (1975) as follows:

$$M_r = \frac{8}{\pi^2} \sum_{n=0}^{\infty} \frac{1}{(2n+1)^2} \exp\left[-(2n+1)^2 \frac{\pi^2 D t}{L^2}\right] \tag{2.15}$$

where $M_r = (M - M_e)/(M_o - M_e)$; M is average moisture content at time t (kg/kg d.b.), M_o is initial moisture content, M_e is equilibrium moisture content, D is moisture diffusivity (m²/s), t is time in seconds, and L is plate thickness (m).

The first term of the series in Equation (2.15) would provide a close approximation. All other terms of the series are small fractions and can be neglected; thus, we obtain:

$$M_r = \frac{M - M_e}{M_o - M_e} = \frac{8}{\pi^2} \exp(-\frac{\pi^2 Dt}{L^2})$$ (2.16)

By simple manipulation of Equation (2.16), the moisture content at time t can be expressed as follows:

$$M = \frac{8}{\pi^2}(M_o - M_e)\exp(-\frac{\pi^2 Dt}{L^2}) + M_e$$ (2.17)

Similarly, average fat content in a sample may be estimated following the same approach. Spatial values of moisture and fat contents in the frying product can be obtained by solving the partial differential equation (2.14) with appropriate boundary conditions.

Ateba and Mittal (1994) developed a model to predict combined heat, moisture, and fat transfer in beef meatballs during deep-fat frying. The authors assumed the single layer meatballs undergo two fat transfer periods; namely, the fat absorption and fat desorption periods during deep-fat frying. The basic diffusion equations were used to describe heat, moisture, and fat transfer in the single layer meatball samples during the absorption period, whereas capillary flow was assumed for fat transfer during the desorption period. Moreira et al. (1995) studied the heat and mass transfer in tortilla chips during deep-fat frying. Single layer models can be used as an initial approximate description of the frying process. Breaded products are inherently composite materials and require models that are more complicated than single layer models.

2.7.2 Composite Layer Models

Ngadi et al. (1997) proposed a model for moisture transfer in chicken drumsticks during deep-fat frying using the finite element method. Chemkhi et al. (2004) proposed a mathematical model for simulating the drying process of potatoes. Yamsaengsung and Moreira (2002a,b) proposed the transport model for tortilla chips during the frying and cooling processes, which also explained various structural changes. Tangduangdee et al. (2003) modeled one-dimensional heat and mass transfer for frozen chicken breasts. Farkas et al. (1996a,b) differentiated the crust and core regions in fried non-breaded potato slices. The authors developed a moving boundary to predict heat and mass transfer processes during frying. Other authors, including Ngadi and Correia (1995) and Ateba and Mittal (1994), have used two-phase models to predict heat and mass transfer in the core and crust parts of non-breaded products. Wang (2005) applied a two-phase model approach to account for simultaneous heat and moisture transfer in both the breading coating and the chicken core during deep-fat frying. The assumptions that were made to formulate the model include:

1. The chicken nugget sample was assumed to be rectangular in shape with no shrinkage during frying.
2. The thickness and width of the chicken nugget was small compared with the length; therefore, the heat and mass fluxes along the length direction were ignored.
3. The heat transfer effects were the same for the top and bottom surfaces; the chicken nugget was symmetric with respect to the heat transfer direction.
4. Boundary conditions for moisture transfer assumed that surface moisture was zero.
5. A moving boundary was assumed.
6. Convective heat transfer coefficient was a constant during the deep-fat frying process.
7. Thermal and mass transfer properties are homogenous within the crust, coating, and core layers but may vary from one layer to the other.

A conceptual schematic of heat and mass transfer in the chicken nuggets is shown in Figure 2.3. A quarter symmetric sample model was partitioned into 330 triangle elements with 187 nodes (Figure 2.4) using the automatic mesh generation of the FEMLAB® software with the maximum element size of 1/15 of the maximum axis parallel distance in the sample geometry and the element growth rate as 1.3 (the maximum rate at which the element size can grow 30% from a region with small elements to a region with larger elements).

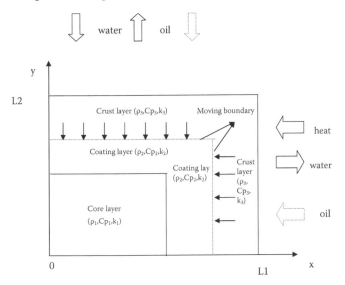

FIGURE 2.3 Schematic of model (half size of chicken nugget both in x and y direction because of symmetry) for heat and mass transfer in coated chicken nugget, where ρ_1, ρ_2, ρ_3, are densities (kg/m³) of the core, coating, and crust layer, respectively. c_{p1}, c_{p2}, c_{p3} are specific heats (kJ/[kg°C]) of the core, coating, and crust layer, respectively. k_1, k_2, k_3, are thermal conductivities (W/mK) of the core, coating, and crust layer, respectively.

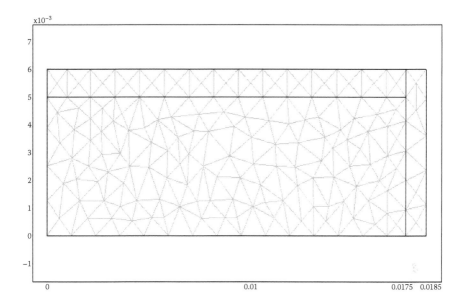

FIGURE 2.4   Schematic diagram of mesh of chicken nugget model. Dimensions are in meters.

The governing equation for rectangular shape (x direction) with phase change is given as:

$$\rho c_p \frac{\partial T}{\partial t} = k_x \frac{\partial^2 T}{\partial x^2} + k_y \frac{\partial^2 T}{\partial y^2} + I \qquad\qquad coating\ part \qquad\qquad (2.18)$$

where k_x and k_y are the thermal conductivities in the x and y directions, respectively. We assumed $k_x = k_y = k$ and I is the rate of internal evaporation heat. Tong and Lund (1993) proposed Equation (2.19):

$$I = \lambda \rho D_e \left(\frac{\partial^2 M}{\partial x^2} + \frac{\partial^2 M}{\partial y^2} \right) \qquad\qquad core\ part \qquad\qquad (2.19)$$

where λ is latent heat (J/kg) of water, D_e is effective diffusivity (m²/s), and M is moisture content (kg/kg).

Boundary conditions (B.C.) are as follows:

$$k \nabla T = 0 \quad x = 0 \quad or \quad y = 0 \qquad\qquad (2.20)$$

$$h(T_f - T_s) = k \nabla T \quad x = L1 \quad or \quad y = L2 \qquad\qquad (2.21)$$

where k is thermal conductivity (W/mK), T_f is frying oil temperature (°C), and T_s is surface temperature of the nugget (°C).

Initial conditions (I.C.) are as follows:

$$T = T_i \tag{2.22}$$

where T_i is the initial temperature of the nugget (°C).

Since the temperature distribution is a function of both space and time, we will assume that the distribution of T within each element has the form

$$T^{(e)}(x,y,t) = \sum_{i=1}^{r} N_i(x,y)T_i(t) \tag{2.23}$$

Applying Galerkin's method for a finite element formulation (Figure 2.4),

$$\iint_{\Omega^{(e)}} N_i[\frac{\partial}{\partial x}(k_x \frac{\partial T^{(e)}}{\partial x}) + \frac{\partial}{\partial y}(k_y \frac{\partial T^{(e)}}{\partial y}) + I - \rho c_p \frac{\partial T}{\partial t}]dxdy = 0 \tag{2.24}$$

After integrating by parts, the matrix equation is obtained.

$$\int_{\Omega^{(e)}} N_i \rho c \frac{\partial T^{(e)}}{\partial t} dxdy = -\iint_{\Omega^{(e)}} (k_x \frac{\partial T^{(e)}}{\partial x}\frac{\partial N_i}{\partial x} + k_y \frac{\partial T^{(e)}}{\partial y}\frac{\partial N_i}{\partial y})dxdy \tag{2.25}$$

$$+ \iint_{\Omega^{(e)}} N_i I\, dx\, dy + \int_{S^{(e)}} (k_x \frac{\partial T^{(e)}}{\partial x}n_x + k_y \frac{\partial T^{(e)}}{\partial y}n_y)N_i d\sum^{(e)}$$

When Equation (2.6) is substituted, the resulting equation for the entire element equation is the following:

$$[K_{c_{ij}}]^{(e)}\{T\}^{(e)} + [C_{ij}]\{\frac{dT}{dt}\}^{(e)} = \{I_i\}^{(e)} - \{q_i\}^{(e)} - [K_{h_{ij}}]^{(e)}\{T\}^{(e)} + \{q_{T_f}\}^{(e)} \tag{2.26}$$

where the superscript (e) restricts the range to one element.

$$K_{c_{ij}} = \iint_{\Omega^{(e)}} (k_x \frac{\partial N_i}{\partial x}\frac{\partial N_j}{\partial x} + k_y \frac{\partial N_i}{\partial y}\frac{\partial N_j}{\partial y})dxdy \tag{2.27}$$

$$C_{ij} = \int_{\Omega^{(e)}} \rho c N_i N_j dxdy \tag{2.28}$$

$$I_i = \iint_{\Omega^{(e)}} I N_i dxdy \tag{2.29}$$

$$q_i = \int_{S^{(e)}} q N_i d \sum\nolimits^{(e)}$$ (2.30)

$$K_{h_{ij}} = \int_{S_2^{(e)}} h N_i N_j d \sum\nolimits^{(e)}$$ (2.31)

$$q_{T_{f_i}} = \int_{S^{(e)}} h T_f N_i d \sum\nolimits^{(e)}$$ (2.32)

Considering the moving boundary located between the coating layer and the crust layer (shown in Figure 2.5), when the temperature at a location inside the coating layer attains the boiling point, the water will change into vapor and evaporate at the surface while this location will become the crust. The boundary will move from the crust to the inner layer (Figure 2.3). Thus, this location's parameters of the properties will change as

$$\rho_2, k_2, c_{p2} \rightarrow \rho_3, k_3, c_{p3}$$

After a long time, the moving boundary will enter the core part; therefore,

$$\rho_1, k_1, c_{p1} \rightarrow \rho_3, k_3, c_{p3}$$

In numerical modeling, discontinuous parameters can lead to problems for the solver. Therefore, in FEMLAB®, ρ, c_p, and k were represented with a smooth function $Y = flc1hs(X, SCALE)$. The function $Y = flc1hs(X, SCALE)$ is a smoothed heaviside function with a continuous first derivative (FEMLAB® 3.0 Reference).

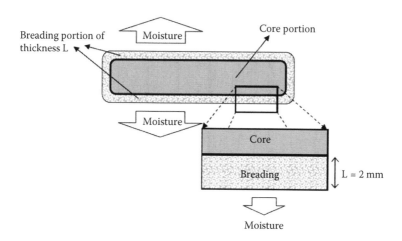

FIGURE 2.5   Schematic drawing of chicken nugget sample.

Heaviside function is a discontinuous step function, usually defined as $h(x) = 0$ for $x < 0$, and as $h(x) = 1$ for $x \geq 0$. $Y = flc1hs(X,SCALE)$. It approximates the logical expression $Y = (X > 0)$ by smoothing the transition within the interval $-SCALE < X < SCALE$ (FEMLAB® Support Knowledge Base). With the help of the function $Y = flc1hs(X,SCALE)$, ρ_2, k_2, and c_{p2} can be changed into ρ_3, k_3, and c_{p3} smoothly. In this case, $Y = flc1hs(T_{trans} - T, SCALE)$, T is the sample temperature, T_{trans} is the transition temperature of water from liquid to vapor, and $SCALE = 1°C$ (FEMLAB® 3.0 Reference). This means that when $-1°C < T_{trans} - T < 1°C$, ρ_2, k_2, and c_{p2} can be transferred to ρ_3, k_3, and c_{p3} smoothly.

2.7.2.1 Modeling Moisture Transfer

Fick's equation was used to model moisture diffusion in the chicken nugget samples:

$$\frac{\partial M}{\partial t} = D_m \left(\frac{\partial^2 M}{\partial x^2} + \frac{\partial^2 M}{\partial y^2} \right) \tag{2.33}$$

where D_m is moisture diffusivity (m²/s) and M is moisture content (kg/kg).
  B.C.:

$$M(L1, y, t) = M(x, L2, t) = \text{constant} = 0 \tag{2.34}$$

where $M(L1, y, t)$ and $M(x, L2, t)$ are the moisture contents of the sample surface (kg/kg) at any frying time.
  I.C.:

$$M(x, y, 0) = M_i \tag{2.35}$$

where $M(x, y, 0)$ is the initial moisture content at any location of the sample (kg/kg) and M_i is the initial moisture content (kg/kg).

  Since during the actual frying operation, many factors may affect the moisture diffusivity, such as pore development during frying, an effective moisture diffusivity, D_{em}, was used in place of moisture diffusivity, D_m, and Fick's equation becomes:

$$\frac{\partial M}{\partial t} = D_{em} \left(\frac{\partial^2 M}{\partial x^2} + \frac{\partial^2 M}{\partial y^2} \right) \tag{2.36}$$

  The finite element equation for moisture transfer was obtained using the Galerkin method, similar to the heat conduction method. Rice and Gamble (1989) reported effective moisture diffusivity to be in the range of 1.40E-09 to 1.55E-08 m²/s for frying potatoes at 145 to 185°C. Furthermore, the effective moisture diffusivity in chicken drumsticks was found to be between 1.32E-09 and 1.64E-08 m²/s during deep-fat frying (Ngadi et al., 1997).

  Average moisture content of a given region was computed by a mass averaged integration of moisture as shown in Equation (2.37):

$$\bar{M} = \frac{\int_S M \, dS}{\int_S dS}$$ (2.37)

where $\bar{M}$ is average moisture content, M is moisture content, and S is area of the sample domain.

2.7.2.2 Modeling Fat Transfer

Similar to moisture transfer, Fick's equation was also used to describe the process of fat transfer during deep-fat frying. Ngadi et al. (1997) employed Fick's model adapted to fat absorption as follows:

$$\frac{\partial M_f}{\partial t} = D_{ef}\left(\frac{\partial^2 M_f}{\partial x^2} + \frac{\partial^2 M_f}{\partial y^2}\right)$$ (2.38)

where D_{ef} is effective oil diffusivity(m²/s) and M_f is the oil content (kg/kg).

The finite element equation for moisture transfer was obtained using the Galerkin method, similar to the heat conduction method. Kassama (2003) reported effective oil diffusivities for chicken breast meat as 9.12E-9, 1.65E-8, and 3.32E-8 m²/s when frying was conducted in 170, 180, and 190°C oil, respectively.

Average oil content could be computed by integration based on the previously mentioned oil transfer model:

$$\bar{M}_f = \frac{\int_s M_f \, ds}{\int_s ds}$$ (2.39)

where $\bar{M}_f$ is average oil content (kg/kg), M_f is oil content (kg/kg), and s is area of the sample domain (m²).

Equations 2.18, 2.33, and 2.38, along with their appropriate boundary conditions, are the coupled 2-D equations that describe heat and mass transfer in breaded chicken nuggets during deep-fat frying. The proposed heat and mass transfer models were solved numerically by the finite element method software (FEMLAB® 3.0). They were used to simulate the center temperature and mass transfer characteristics of coated chicken nuggets.

The authors reported that coating thickness and pretreatment of the product influenced heat and mass transfer characteristics. Thus, the formulation of the breading coating as well as its thickness (i.e., pickup) are important product parameters of consideration in designing optimized fried breaded products.

In a study conducted by Moreira et al. (1995), an attempt was made to model the simultaneous heat and mass transfer in the frying of tortilla chips. The equations that were produced were solved using the finite difference method. The finite difference method is commonly used in the modeling process and can be a powerful tool in predicting certain parameters involved in the frying process. However, in most cases, the equations were only valid during certain periods in the process. For instance, some models may only be good predictors during the first five minutes of frying. Perhaps after five minutes, one of the factors that were assumed to be negligible comes into play and has an effect on the frying process.

A considerable amount of work done on deep-fat frying thus far has been primarily limited to non-fatty foods (such as potatoes) and the mechanism of fat transport in products containing initial fat has not been thoroughly researched. An understanding of fat transfer during frying of meat products is of interest to processors, as consumers become more concerned about the amount of fat they consume (Ateba and Mittal, 1994). To capture a good picture of mass transport in fried foods generally with the aim of improving the quality, and understanding of the structured changes and how they relate to mass transfer during frying is also very important (Aguilera and Stanley, 1999).

2.8 REFERENCES

Aguilera, J.M. and Stanley, D.W. 1999. *Microstructural principles of food processing and engineering.* Gaithersburg, MD: Aspen Publishers. pp. 309–310.

Albert, S. and Mittal, G.S. 2002. Comparative evaluation of edible coatings to reduce fat uptake in a deep-fried cereal product. *Food Res. Intl.* 35(5): 445–458.

Ashkenazi, N., Mizrahi, S., and Berg, Z. 1984. Heat and mass transfer in frying. In *Engineering and foods,* Volume 1, McKeena, B.M. (Ed.), London: Elsevier Applied Science, pp. 109–116.

Ateba, P. and Mittal, G.S. 1994. Dynamics of crust formation and kinetics of quality changes during frying of meatballs. *J. Food Sci.* 59(6): 1275–1278.

Baker, R., Baldwin, E., and Nisperos-Carriedo M. 1994. Edible coatings and films for processed foods. In *Edible coatings and films to improve food quality*, J.M. Krochta, E.A. Baldwin, and M.O. Nisperosarriedo (Eds.), Technomic Publishing Co. Inc., Lancaster, pp. 89–104.

Balasubramaniam,V.M., Chinnan, M.S., and Mallikarjunan, P. (1995). Deep-fat frying of edible film coated products: Experimental and modelin. In Proceedings of the FPAC IV Condeference. St. Joseph. MI: American Association of Agricultural Engineers, pp. 486–494.

Balasubramaniam, V.M., Chinnan, M.S., Mallikarjunan, P., and Phillips, R.D. 1997. The effect of edible film on oil uptake and moisture retention of a deep-fat fried poultry product. *J. Food Proc. Eng.* 20(1): 17–29.

Blumenthal, M.M. 1991. A new look at the chemistry and physics of deep fat frying. *Food Technol.* 45(2): 68–72, 94.

Budzaki, S. and Seruga, B. 2005. Determination of convective heat transfer coefficient during frying of potato dough. *J. Food Eng.* 66(3): 307–314.

Chemkhi, S., Zagrouba, F., and Bellagi, A. 2004. Mathematical model for drying of highly shrinkable media. *Drying Technol.* 22(5): 1023–1039.

Costa, R.M., Oliveira, F.A.R., Delaney, O., and Gekas, V. 1999. Analysis of the heat transfer coefficient during potato frying. *J. Food Eng.* 39(3): 293–299.

Crank, J. 1975. *The arithmetics of diffusion,* 2nd ed. Oxford, U.K.: Oxford University Press.

Davis, A. 1983. Batter and breading ingredients. In *Batter and breading technology,* Suderman, D.R. and Cunningham, F.E. (Eds.), Westport, CT: AVI Publishing Company, p. 15.

Debeaufort, F., Quezada-Gallo, J.A., and Voilley, A. 1998. Edible films and coatings: Tomorrow's packagings: A review. *Crit. Rev. Food Sci. Nutr.* 38(4): 299–313.

Dincer, I. 1996. Modeling for heat and mass transfer parameters in deep frying of products. *Heat Mass Transfer* 32: 109–113.

Dow Chemical Co., Fried foods stabilizer keeps moisture in, fat out. Food development. *Prepared Foods*, March (1991), pp. 61.

Dziezak, J.D. 1989. Fats, oils, and fat substitutes. *Food Technol.* 43: 66–74.

Farkas, B.E. (1994). Modeling immersion frying as a moving boundary problem. Unpublished Ph.D. Dissertation, University of California, Davis.

Farkas, B.E., Singh, R.P., and Rumsey, T.R. 1996a. Modeling heat and mass transfer in immersion frying. I. Model development. *J. Food Eng.* 29(2): 211–226.

Farkas, B.E., Singh, R.P., and Rumsey, T.R. 1996b. Modeling heat and mass transfer in immersion frying. II. Model solution and verification. *J. Food Eng.* 29(2): 227–248.

Flores, S., Famá, L., Rojas, A.M., Goyanes, S., and Gerschenson, L.N. 2007. Physical properties of tapioca-starch edible films: Influence of filmmaking and potassium sorbate. *Food Res. Intl.* 40(2), 257–265.

Gamble, M.H. and Rice, P. 1988. Effect of initial tuber solids content on final oil content of potato chips. *Lebensmittel-Wissenschaft und-Technologie* 21: 62–65.

Gamble, M.H. and Rice, P. 1987. Effect of pre-fry drying of oil uptake and distribution in potato crisp manufacture. *Intl. J. Food Sci. Technol.* 22(5): 535–548.

Gamble, M.H., Rice, P., and Selman, J.D. 1987. Relationship between oil uptake and moisture loss during frying of potato slices from c. v. Record U.K. tubers. *Intl. J. Food Sci. Technol.* 22(3): 233–241.

Guillaumin, R. (1988). Kinetics of fat penetration in food. In *Frying of food*, Valera, A.E., Bender, I.D., and Merlon (Eds.), Chichester, U.K.: VCH.

Hallstrom, B., Skjoldebrand, C., and Tragardh, C. 1988. *Heat transfer and food products*. New York: Elsevier Applied Science.

Indira, T.N., Latha, R.B., and Prakash, M. 1999. Kinetics of deep-fat-frying of a composite product. *J. Food Sci. Technol.* 36(4): 310–315.

Innawong, B. 2001. Improving fried product and frying oil quality using nitrogen gas in a pressure frying system. Unpublished Ph.D. Dissertation, Virginia Polytechnic Institute and State University.

Kassama, L.S. 2003. Pore development in food during deep-fat frying. Macdonald Campus, McGill University, Quebec, Canada. PhD thesis.

Kester, J.J. and Fennema, O.W. 1986. Edible films and coatings: A review. *Food Technol.* 40, 47–59.

Krochta, J.M. and de Mulder-Johnston, C. 1997. Edible and biodegradable polymer films: Challenges and opportunities. *Food Technol.* 51: 60–74.

Krokida, M.K., Oreopoulou, V., and Maroulis, Z.B. 2000. Water loss and oil uptake as a function of frying time. *J. Food Eng.* 44(1): 39–46.

Krokida, M.K., Oreopoulou, V., Maroulis, Z.B., and Marinos-Kouris, D. 2001. Deep fat frying of potato strips-quality issues. *Drying Technol.* 19(5): 879–935.

Lane, R.H., Muir, W.M., and Mullins, S.G. 1980. The influence of fryer temperature and raw weight on fry time of deep fat fried chicken thighs. *Poultry Sci.* 59: 2467–2469.

Levine, L. 1990. Understanding frying operations, part I. *Cereal Foods World* 35(2): 272.

Loewe, R. 1990. Ingredient selection for batter systems. In *Batters and breadings in food processing,* Kulp, K. and Loewe, R. (Eds.), St. Paul, MN: American Association of Cereal Chemists, pp. 11–28.

Magnuson, K.M. (1985). Uses and functionality of vital wheat gluten. *Cereal Foods World* 30(2): 179–180.

Mallikarjunan, P., Chinnan, M.S., Balasubramaniam, V.M., and Phillips, R.D. 1997. Edible coatings for deep-fat frying of starchy products. *Lebensmittel-Wissenschaft und-Technologie*, 30(7): 709–714.

Mallikarjunan, P., Chinnan, M.S., and Balasubramaniam, V.M. 1995. Modeling deep-fat fry-ing of chicken coated with edible films. Paper presented at the IFT Annual Meeting., Paper No. 87-2.

Mate, J.I. and Krochta, J.M. 1996. Comparison of oxygen and water vapor permeabilities of whey protein isolate and β-lactoglobulin edible films. *J. Agric. Food Chem.* 44(10): 3001–3004.

Miller, K.S., Singh, R.P., and Farkas, B.E. 1994. Viscosity and heat transfer coefficients for canola, corn, palm, and soybean oil. *J. Food Proc. Preserv.*, 18(6): 461–472.

Mittelman, N., Mizrahi, S., and Berk, Z. (1982). Heat and mass transfer in frying. In *Engineering and foods,* McKeena, B.M. (Ed.), London: Elsevier Applied Science, pp. 109–116.

Mohsenin, N.N. 1980. Thermal Properties of Foods and Agricultural Materials. New York: Gordon and Breach.

Moreira, R.G., Palau, J., Castell-Perez, M.E., and Sweat, V.E. 1991. Moisture loss and oil absorption during deep-fat frying of tortilla chips. Presented at Winter Meeting., American Society of Agricultural Engineers, St. Joseph, Michigan.

Moreira, R., Palau, J., and Sun, X. 1995. Simultaneous heat and mass transfer during deep fat frying of tortilla chips. *J. Food Proc. Eng.* 18(3): 307–320.

Mukprasirt, A., Herald, T.J., and Flores, R.A. 2000. Rheological characterization of rice flour-based batters. *J. Food Sci.* 65(7): 1194–1199.

Ngadi, M., Dirani, K., and Oluka, S. (2006). Mass transfer characteristics of chicken nuggets. *Intl. J. Food Eng.*, 2(3): 8, 1–16.

Ngadi, M. and Ikediala, J.N. 1998. Heat transfer properties of chicken-drum muscle. *J. Sci. Food Agric.* 78: 12–18.

Ngadi, M. and Ikediala, J.N. (2005). Natural heat transfer coefficients of chicken drum shaped bodies. *Intl. J. Food Eng.* 1(3): 4.

Ngadi, M.O. and Correia, L.R. 1994. Kinetic modelling of chicken muscle thermal conductiv-ity during deep-fat frying, In *Developments in food engineering*, T. Yano, R. Matsuno and K. Nakamura (Eds.), 513–515. Blackie academic and Professional, New York.

Ngadi, M.O. and Correia, L.R. 1995. Moisture diffusivity in chicken drum muscle during deep-fat frying. *Canadian Agric. Eng.* 37(4): 339–344.

Ngadi, M.O., Watts, K.C., and Correia, L.R. 1994. Specific heat capacity of chicken drum muscle. Paper No. 71B-24 presented at the annual meeting of the Institute of Food Technologists (IFT). Atlanta, Georgia.

Ngadi, M.O., Watts, K.C., and Correia, L.R. 1997. Finite element method modeling of mois-ture transfer in chicken drum during deep-fat frying. *J. Food Eng.* 32(1): 11–20.

Ni, H. and Datta, A.K. 1999. Moisture, oil and energy transfer port during deep fat frying food materials. *Transacts. Inst. Chem. Eng.* 77(2): 194–203.

Nisperos-Carrieedo, M.O. (1994). Edible coatings and films based on polysaccharides. In *Edible coatings and films to improve food quality,* Krochta, J.M., Baldwin, E.A., and Nisperos-Carriedo, M.O. (Eds.), Lancaster, PA: Technomic Publishing Company, pp. 305–335.

Pannu, K.S. and Chinnan, M.S. 1999. Heat transfer coefficients obtained with metal balls during food frying. *Proc. 6th Conf. Food Eng. (CoFE '99)*, Dallas, TX. Oct. 31–Nov. 5, pp. 53–58.

Park, H.J. and Chinnan, M.S. 1995. Gas and water vapor barrier properties of edible films from protein and cellulosic materials. *J. Food Eng.* 25(4): 497–507.

Perkins, E.G. and Erickson, M.D. 1996. Regulation of frying fat and oil in deep frying. In *Chemistry, nutrition, and practical application.* Champaign, IL: ACCS Press, pp. 323–334.

Pinthus, E.J., Weinberg, P., and Saguy, I.S. 1995. Oil uptake in deep fat frying as affected by porosity. *J. Food Sci.* 60(4): 767–769.

Pinthus, E.J. and Saguy, I.S. 1994. Initial interfacial tension and oil uptake by deep-fat fried foods. *J. Food Sci.* 59(4): 804–807.

Rao, M.A., Rizvi, S.S.H., and Datta A.K. 2005. Engineering Properties of Foods. New York: Marcel Dekker.

Rice, P. and Gamble, M. 1989. Modeling moisture loss during potato slice frying. *Intl. J. Food Sci. Technol.* 24(2): 183–187.

Saguy, I.S. and Pinthus, E.J. 1995. Oil uptake during deep-fat frying: Factors and mechanism. *Food Technol.* 49(4): 142–145,152.

Sahin, S., Sastry, S.K., and Bayindirli, L. 1999. Heat transfer during frying of potato slices. *Lebensmittel-Wissenschaft und-Technologie* 32(1): 19–24.

Sanderson, G.R. 1981. Polysaccharides in food. *Food Technol.* 35: 50–56, 83.

Sanz, M.D. Alonso and R.H. Mascheroni. Thermophysical properties of meat products: General bibliography and experimental values. *Transactions of ASAE* 30 (1): 283–289.

Singh, P.R. 1995. Heat and mass transfer in foods during deep-fat frying: Engineering aspects of deep-fat frying of foods. *Food Technol.* 49(4): 134–137.

Skjoldebrand, C. and Olsson, H. 1980. Crust formation during frying of minced meat product. *J. Food Sci. Technol.* 13: 148.

Sosa-Morales, M.E., Orzuna-Espiritu, R., and Vélez-Ruiz, J.F. 2006. Mass, thermal and quality aspects of deep-fat frying of pork meat. *Journal of Food Engineering* 77: 731–738.

Suderman, D.R. (1983). Use of batters and breadings on food products: A review. In *Batter and breading technology,* Suderman, D.R. and Cunningham, F.E. (Eds.), Westport, CT: AVI Publishing Company, pp. 1–13.

Taiwo, K.A. and Baik, O.D. 2007. Effects of pre-treatments on the shrinkage and textural properties of fried sweet potatoes. *Lebensmittel-Wissenschaft und-Technologie* 40(4): 661–668.

Tangduangdee, C., Sakarmdr, B., and Suvit, T. 2003. Heat and mass transfer during deep-fat frying of frozen composite food with thermal protein denaturation as quality index. *Sci. Asia* 29: 355–364.

Tong, C.H. and Lund, D.B. 1993. Microwave heating of baked dough products with simultaneous heat and moisture transfer. *J. Food Eng.* 19: 319–339.

Tseng, Y.C., Moreira, R.G., and Sun, X. 1996. Total frying-use time effects on soybean-oil deterioration and on tortilla chip quality. *Intl. J. Food Sci. Technol.* 31(3): 287–294.

Wang, Y. 2005. Heat and mass transfer in deep fat frying of breaded chicken nuggets. McGill University, Montreal.

White, P.J. 1991. Methods for measuring changes in deep-fat frying oils. *Food Technology* 45: 75–83.

Yamsaengsung, R. and Moreira, R.G. 2002a. Modeling the transport phenomena and structural changes during deep fat frying: Part I: model development. *J. Food Eng.* 53(1): 1–10.

Yamsaengsung, R. and Moreira, R.G. 2002b. Modeling the transport phenomena and structural changes during deep fat frying: Part II: model solution and validation. *J. Food Eng.* 53(1): 11–25.

Yildiz, A., Palazoglu, K., and Erdogdu, F. 2007. Determination of heat and mass transfer parameters during frying of potato slices. *J. Food Eng.* 79(1): 11–17.

3 Fryer Technology

The frying process results in food dehydration leading to a low activity shelf-stable product. As a result, manufacture of fried foods has evolved into a very large segment of the food industry, a multibillion industry in the United States and abroad. Although the frying process is very simple in principle, it becomes very sophisticated and complex as the scale of production increases from home-scale processing to restaurant-type cooking to very large scale industrial processing. For the processors to be competitive, it is extremely important to have efficient frying equipment that is capable of meeting the high demands of deep-fat fried foods. There have been increased efforts to improve upon the design of fryers. In the past 30 years, there has been a significant increase in the variety of deep-fat fryers available on the market. In fact, there are more different types of frying equipment than any other heat transfer procedure in food service (Cummings, 1983). The goal of each new design is to provide high-quality fried foods while maintaining the quality of the frying oil. An effective automated frying system consists of four requirements: (1) accurate temperature control, (2) efficient heat transfer, (3) minimal oil contamination, and (4) minimal oil turnover (Moreira, Castell-Preez, and Barrufet, 1999).

The choice of which fryer type to use is critical because the type of fryer greatly influences the quality of the final product. The product characteristics such as physical dimensions, the presence or absence of a coating, and the desired sensory qualities must be considered when selecting a fryer type. Other factors to be considered include cost, space, and safety.

Frying can be accomplished in a batch or in a continuous system. Batch fryers are typically smaller and are used primarily in catering-type restaurants. Batch frying capabilities are limited, making the system impractical for large quantity production (Padilla, 1998). Continuous fryers that are capable of handling large amounts of frying oil and food materials are primarily used in industrial settings, which involve large-scale production. The fryers can be operated under atmospheric, low (even vacuum), or high pressure. In addition, it is also possible to perform frying under inert gas atmosphere, which has the specific advantages of reduced oxidation of frying oil and better-quality fried product (Innawong, Mallikarjunan, and Cundiff, 2006). Most large-scale production continuous frying is done under atmospheric conditions. There are two basic means of heating the oil within the fryer, referred to as direct and indirect heating systems. The oil may be heated outside the fryer, which is appropriately called an external heating system. All heating systems have their advantages as well as their disadvantages. The decision about the type of frying system to be used is critical to the quality of the finished product and overall efficiency of the frying process.

3.1   BASIC COMPONENTS OF A FRYING SYSTEM

The basic components of a frying system are outlined in Table 3.1. The components are similar to both continuous and batch fryers (Stier, 1996).

TABLE 3.1

Basic Components of a Frying System

Oil holding kettle	Most operations add fresh oil to the fryer to ensure proper oil level; however, some operations have this provision for the source of oil.
Conveying system	As the name implies, it conveys the product into and out of the fryer. In continuous fryers, the conveyors also move the product through the fryer. The type is product dependent.
Heating and control system	Heats and controls the oil temperature. These are described in the heating systems section.
Canopy	Also called a hood. It covers the fryer (in case of continuous systems) to conserve energy. The ducts and blowers installed in the canopy remove the steam and volatiles.
Hoist	Mechanically or manually lifts the canopy (hood) and conveyor from the fryer.
Filter system	A feature on all fryers to remove the solids — sinkers, floaters, or the in-between. It may be a passive (a simple metal screen with or without filter paper) or an active (using filter aid or chemical adsorbent) system.
Frying vessel	The main cavity/well for holding the frying oil in continuous or batch fryers, where the frying process takes place.
Cook area	The available cooking surface area, calculated by multiplying cooking width by cooking length.
Cook time	The time measured from when the product is fully immersed into the oil until it exits or is taken out. In continuous fryers, it is a function of the conveyor, which can be adjusted using variable speed drives.
ΔT	Refers to the temperature differential (°C or °F) for the hot oil during frying and is a most significant parameter in designing external heat exchange systems. An acceptable ΔT is a factor in selecting the fryer best suited for a product to be fried.
Film (skin) temperature	The oil temperature reached at the surface of the heat exchanger tube.
Net heat load (Btu/h)	The quantity of heat required to cook the product not including the heat losses.
Gross heat load (Btu/h)	Net heat load multiplied by the production rate (lb/h) plus system heat losses from convection and radiation.
Oil turnover (h)	Considered the single most important parameter in fryer design. It is the ratio of oil volume or weight in a system to the hourly rate of oil picked up by the product. As an example, if a frying system has a capacity of 5000 lb of oil and the product being fried picks up 500 lb of oil per hour, then the oil turnover is 10 h (calculated by dividing the fryer capacity by oil pick up rate). Product quality is compromised when the oil turnover time is excessive or extremely low.

3.1.1   HEATING SYSTEMS

3.1.1.1   Direct Heating System

In a direct heating system, the frying oil is heated inside the fryer by heating elements such as burner tubes or electric elements that are completely immersed in the oil (Figure 3.1). Heat is conducted from the heating elements to the oil in its immediate vicinity and then the heat is transferred throughout the oil by means of natural convection.

The heating elements are arranged lengthwise, widthwise, or in a single S-shaped configuration. A buffer area of oil between the heat transfer mechanism and the product is required because it allows natural convection currents to rise and have adequate time to heat the oil uniformly by eliminating hot spots in the fryer (Padilla, 1998). The heating capacity of this type of system is limited to the available surface area of the heating elements. Stier (1996) listed several advantages and disadvantages of direct heating systems, which are given in Table 3.2. Direct heating systems are typically used when trying to maximize floor space and when lower costs are desired.

3.1.1.2   Indirect Heating System

In an indirect heating system, a thermal fluid is heated externally with a heater or boiler fired with oil, gas, or electricity. These thermal fluids can be steam but are usually chlorinated hydrocarbons. Steam is the preferred choice of thermal fluid in the United States and chlorinated hydrocarbons are preferred in Europe. The thermal fluid is normally passed through thermal tubes and fins, which are immersed in the oil trough of the fryer. A three-way valve, which regulates the thermal fluid entering the fryer system, controls the temperature. A temperature probe installed in the fryer provides feedback to modulate the three-way valve to release a portion of the thermal fluid into the fryer and the remaining fluid is bypassed by the continuously running

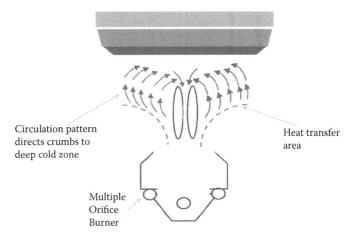

FIGURE 3.1   Direct heating system with a unique fry pot design with deep cold zone for a restaurant type Frymaster™ fryer (Endois, Shreveport, LA).

TABLE 3.2
Advantages and Disadvantages of a Directly Heated Fryer

Advantages	Disadvantages
Compact packages	Lack of overall temperature control
Lower initial capital investment	Relatively high oil volume
Some temperature zoning	Limited heat capacity
Flexibility (multipurpose)	Lower thermal efficiency
	Mechanical removal of fines
	Settlement of fines on heating elements
	Limited depth of product frying zone

Source: Steir, R.F. 1996. Understanding high-volume frying, Part 1. *Baking and Snack*, 18 (2): 70–76.

circulation pump. The schematic in Figure 3.2 illustrates an indirect heating system. Such systems are commonly used for frying battered and breaded foods. The systems using chlorinated hydrocarbons are not commonly used in the United States because it is erroneously believed that the chlorinated hydrocarbons may contaminate the frying oil and, in turn, taint the product fried in the fryer. A list of advantages and disadvantages of an indirectly heated fryer are presented in Table 3.3.

3.1.1.3   External Heating System

External heat exchangers are used to control and maintain the desired temperature of the oil in the fryer (Figure 3.3). The heating source varies and can include gas, fuel oil, or electricity. Feedback temperature from a temperature-sensing probe is used to control the thermal energy into the heat exchanger, which in turn controls the frying oil temperature. The most common heat exchangers used are coil-type heat exchangers and shell and tube exchangers (Padilla, 1998). In external heating systems, oil is circulated through the heat exchanger and returned directly back to the fryer. Through the course of being passed through the heat exchanger, the oil is

TABLE 3.3
Advantages and Disadvantages of an Indirectly Heated Fryer

Advantages	Disadvantages
More uniform temperature control	Limited heat capacity
For gas-fired elements, no gas equipment in the cooking area	Mechanical removal of settling fines on heating elements
Less noisy	Limited product pack depth
High thermal efficiency	Difficulty in cleaning
Temperature zoning possible	Intensive capital requirement
	Large floor space requirements

Source: Steir, R.F. 1996. Understanding high-volume frying, Part 1. *Baking and Snack*, 18 (2): 70–76.

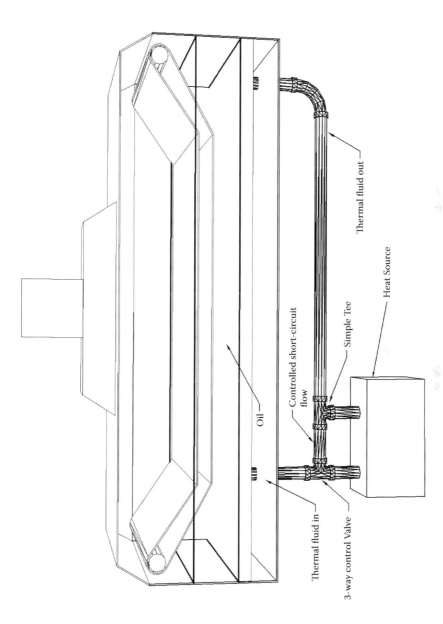

FIGURE 3.2 A schematic diagram of an indirect heating system in an industrial fryer.

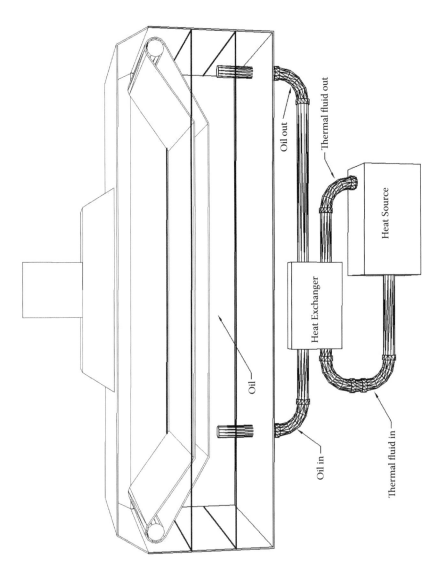

FIGURE 3.3 A schematic diagram of an external heating system in an industrial fryer.

TABLE 3.4

Advantages and Disadvantages of an Externally Heated Fryer

Advantages	Disadvantages
Uniformity of temperature control	Noise from pump/motor
Extremely high heat capacity	Intensive capital requirement
For gas-fired elements, no gas equipment in the cooking area	Large floor space requirements
Low oil volume	
Highest thermal efficiency	
Temperature zoning possible	
Automatic CIP systems usually integrated into the system	
Oil and product velocity is matched	

Source: Steir, R.F. 1996. Understanding high-volume frying, Part 1. *Baking and Snack*, 18 (2): 70–76.

also filtered. That is, excess food particles that cause the oil to degrade are removed. There are many advantages to this type of system. Not only is the oil continuously filtered but the amount of oil required to fry is significantly less than that required in the direct heating system. Because of the lower oil volumes, there is a better oil turnover rate. The main advantages and disadvantages of external heating systems are described in Table 3.4.

3.1.2 FRYER FILTRATION

The removal of debris from frying oil is essential to the quality of fried foods because it alters the taste and appearance of the food material. The products also impart color compounds into the oil, causing it to darken. Various filters and screens can be used to remove the undesirable debris particles. Screens can be used to eliminate larger particles and should be the first step used in the removal process (Jacobson, 1991). The oil should then pass through a filter designed to remove smaller particles. Alkaline filtration systems can also be employed. However, alkaline treatments lead to soap formation, which can be introduced into the oil even after filtration and cause darkening and undesirable flavor changes in the oil (Jacobson, 1991).

3.2 RESTAURANT AND FOOD SERVICE FRYING EQUIPMENT

There are various types of fryers used in the restaurant and fast-food industry. Pictures of some selected batch-type food service fryers are given in Figure 3.4. Most high production frying, especially in quick-service restaurants, is done with floor-mounted models. These fryers are capable of holding from 15 lb to over 200 lb of oil. The oil (frying medium) is typically heated by atmospheric or infrared (IR) burners under the kettle or in the "fire tubes" that pass through the frying medium (Bendall, 1998; Fisher, 2002). A typical baffled heat exchanger for a gas-fired burner is shown in Figure 3.5. Energy input rates range from 30 to 260 kBtu/h, with 35- to 50-lb oil capacity French fryers in the range of 80 to 120 kBtu/h. Innovations in

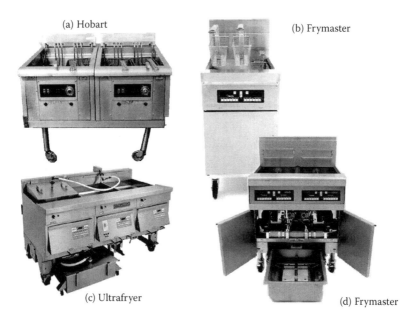

(a) Hobart

(b) Frymaster

(c) Ultrafryer

(d) Frymaster

FIGURE 3.4 Various types of commercial batch-type food service fryers. (a) Twin Hobart Electric Fryer, (b) Frymaster Electric Dual Basket Floor Model, (c) Ultrafryer Systems® Gas PAR-2 S & SC, and (d) Frymaster Gas fryer Model #H255 (shown with optional computers and FootPrint PRO® filter).

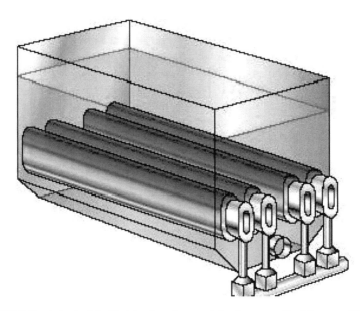

FIGURE 3.5 A typical baffled heat exchanger tube for a gas-fired burner. (Courtesy of Frymaster, Shreveport, LA.)

burner design and application, such as powered infrared burners, have dramatically improved fryer performance. The operating temperatures vary from 160 to 177°C and frying time ranges from 7 to 10 min (Moreira et al., 1999). Some fryers are designed to have very quick response times for a rapid heat recovery in order to meet the high demand in restaurants, especially fast-food service establishments.

Controllers on these units are usually nothing more than an on–off power switch that turns on the power to the heating elements or to the gas control switch. The control system may have two operating modes — standby and fry. The frying unit is kept in standby mode during the idle period when the demand is low; for example, after a rush hour period. During frying when the unit is just loaded, the temperature immediately drops several degrees. Therefore, it is essential that the temperature recovery time be within a specific time range to obtain the desired product characteristics (texture, color, appearance, greasiness-oil absorption, and flavor). Table 3.5 indicates a typical oil temperature profile during frying of par-fried coated chicken in a twin-pan countertop fryer, Model 5301A (Star Manufacturing International, Smithfield, TN).

Advanced gas fryers can consume up to 25% less energy than standard, medium efficiency atmospheric units can and up to 40% less than inexpensive, low efficiency, "throw-away" fryers can. The more heat transferred from the combustion process to the frying medium, the higher the productivity and the lower the energy cost. Figure 3.6 illustrates how an efficient fryer is more desirable than a base-model fryer. As a rule of thumb, the higher the average oil temperature, the better the product quality. With shorter cooking and recovery times, the more efficient fryer is able to produce nearly twice the amount of food as the base-model fryer.

One of the most popular fryers used in the fast food industry is the pressure fryer (Figure 3.7). This fryer type was initially designed for the frying of chicken and has been shown to produce juicier and crispier products than open fryers (Rao and Delaney, 1995; Mallikarjunan, Chinnan, and Balasubramaniam, 1997). By appropriately selecting temperature and pressure in deep-fat frying, consumer desired characteristics could be obtained in the end product (Rao and Delaney, 1995).

TABLE 3.5
Fryer Oil Temperature Profile during Frying of Par-Fried Coated Chicken

Steps	Fryer oil temperature [°F (°C)]
Initial fryer oil temperature	360 ± 2 (182 ± 1)
Rapid temperature drop of the oil in the fryer	330 (166)
Temperature after 4 min	350 (177)
Product interior temperature	160 (71)
Oil temperature after the product was taken out and the temperature allowed to recover for the next batch	360 ± 2 (182 ± 1)

Source: Gupta, M.K., Warner, K., and White, P.J. 2004. *Frying technology and practices*, Urbana, IL: AOCS Press.

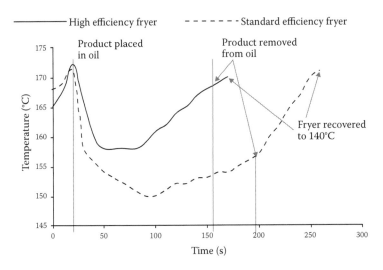

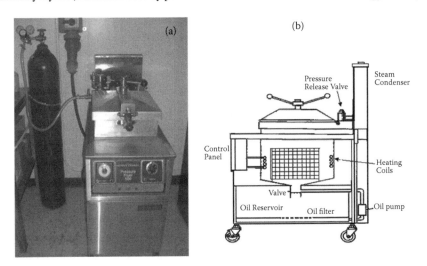

FIGURE 3.6 Fryer oil temperature while cooking a heavy load of products in high- and low-efficiency fryers. (Illustration used by permission of PG & E Food Service Technology Center.)

FIGURE 3.7 Modified restaurant type pressure fryer (Model 500, Henny Penny Inc.) to incorporate external gases for pressure development. (Courtesy of Virginia Tech.)

Pressure batch fryers can be electric or gas-fired and are available in many sizes. In a pressure frying system, a lid covers the fryer so that it is not exposed to the atmosphere. As the food material heats and begins to release moisture in the form of steam, there is a buildup of pressure in the air space above the oil. Once the air space becomes completely saturated with steam, water molecules are no longer able to escape from the food material. This results in a moist, juicy finished product. Although the steam helps to maintain moisture within the product, it also aids in diminishing the quality of the oil. The water creates oxidative reactions in the frying oil, which lead to the formation of undesirable decomposition products.

FIGURE 3.8 A countertop fryer (Model STR 630FD-2) used in the food service industry. (Courtesy of Star Manufacturing International Inc.)

The food service industry also frequently uses countertop fryers because they are relatively inexpensive and require less space (Figure 3.8). In fact, some of these fryers can hold two fry baskets in a 12-in. counter space (Durocher, 1991). The most prevalent countertop units are electric.

Thermal oil degradation is a major issue in food service fryers. This is due to prolonged exposure of oil to high temperatures and frequent idle times during the day when no frying is done but oil temperature is maintained for frying the product on demand. Generally, the oil is filtered twice a day. Some restaurants may do so three times a day. As mentioned before, the filtration system is integrated into the fryer; however, external filtration systems are also available.

3.3 INDUSTRIAL FRYING EQUIPMENT

There are three types of industrial fryers — batch, continuous, and vacuum. Batch fryers are similar to those described previously in food service or restaurant-type units. The primary difference is the capacity. Generally these are not used for large-scale productions but for specialty, or niche market, products (for example, kettle-fried hard-bite potato chips, also called home-style chips).

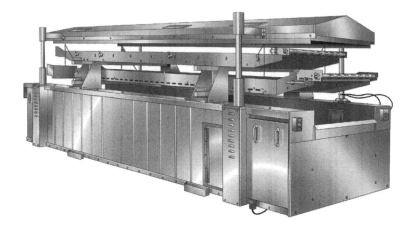

FIGURE 3.9 A continuous industrial fryer, Stein HPF11 High Performance Fryer. (Courtesy of JBT Food Tech, Chicago, IL) This fryer is 23 ft long, has a 34 in. wide belt, is gas fired, and has a heating capacity of 3.5 million BTU.

Continuous fryers are designed for large-scale production (Figure 3.9). The unit shown in the figure is a direct-heat gas fired continuous fryer specifically designed for high volume processing requirements. It also features an insulated hood and an automatic hood-lift system that raises hood and conveyors for maintenance and easy sanitation access. The internal design and construction varies with the type of product to be fried. Vacuum fryers are expensive and have high operational costs, but they do not have high production capacity. The concept of continuous vacuum frying was first introduced by Florigo of Holland (H&H Industry Systems B.V., The Netherlands) for producing high-quality French fries. Today these are used for high-end market products for frying fruits and vegetables; such fryers are not typically used for breaded and coated products. According to Moriera et al. (1999), in the current environment of consumers' preoccupation with lower fat products, vacuum frying would provide a potential alternative of reducing oil content in fried products. Several characteristics of vacuum fryers are listed (Gupta, Warner, and White, 2004):

1. Typical operating pressure is less than 100 mm Hg.
2. Frying is conducted at 250°F, a much lower temperature than atmospheric frying.
3. The food is placed in the basket and the basket is inserted and placed in the vacuum chamber above the oil surface.
4. The vacuum is applied and the basket is lowered into the oil.
5. Frying begins and the oil temperature drops as in batch fryers.
6. The oil is circulated continuously through an external heater.

7. The oil regains the temperature as the desired moisture level in the product is achieved.
8. The vacuum is broken gently, the fryer is opened, and the product basket is lifted out of the oil.
9. The excess oil is allowed to drain. Then the product is cooled and is ready to be packaged.

3.4 CRITICAL FACTORS IN SELECTION OF FRYING EQUIPMENT

The fryer must meet the processing protocol and conditions for producing a desirable product. Gupta et al. (2004) provided a list of items to be considered in the selection of a fryer:

Knowledge of desired product characteristics
Production volume (fryer capacity)
Heat load estimation
Conveyor system
Oil turnover rate
Debris and fines removal
Cleaning and maintenance protocols
Emission requirements
Heating system
Technical support from the fryer manufacturer/supplier

In this book, the emphasis is on frying of battered and breaded foods so the desired food characteristics and coating systems (breading, battering, dusting, etc.) specific to that focus are discussed in Chapter 5. Here we focus only on the fryer capacity, heat load requirements, conveyors, Delta-t, oil turnover time, fryer oil treatment, emissions, and cleaning and maintenance.

3.4.1 FRYER CAPACITY

This is selected based on the quantity of the product to be fried as per the requirement of the enterprise. A number of factors are taken into account for estimating frying capacity. They include: (1) the operation hours of the business per day — one shift (8 h), two shifts (16 h), three shifts (24 h), or in between; (2) cook time for the product to achieve desired product characteristics (moisture, appearance, texture, and flavor); (3) fryer maintenance and sanitation protocol; and (4) product shelf life and company policy for warehousing, shipping, and distribution.

This information is instrumental for estimating the physical dimensions of the fryer, which include the conveyor width and length, and the loading rate. More specifically, this is defined as the number of pieces per linear foot of the conveyor; pound of product per linear foot of the conveyor; and product (lb) per unit area (ft^2)

of the conveyor. An example for estimating the fryer loading (lb/ft^2) is presented as follows:

Desired frying capacity   = 4500 lb/h = 75 lb/min
System or conveyor loading rate   = 2 lb/ft^2
Required cooking surface per min (frying capacity/loading rate)   = 37.5 ft^2/min

If the fry time (process time) is 2 min:

Fryer/cook surface area will be (37.5 ft^2/min * 2 min)   = 75 ft^2

If the width is selected as 2.5 ft:

The length will be calculated as (75 ft^2/ 2.5 ft)   = 30.0 ft

Changing the frying capacity and fry time will change these requirements. Alternatively, if the same fryer is used for a product with a different fry time (say 3 min), then the frying capacity will reduce to 3000 lb/h.

3.4.2 HEAT LOAD

Cooking in the fryer is primarily a dehydration process. The thermal energy in the hot oil is transferred to the product being cooked, causing the moisture in the product to evaporate. This implies the energy is needed for the heat of vaporization as well as for conduction heating of the product because there is a temperature differential between the product entering the hot oil and the oil temperature. In addition, there is energy loss at various points in the system. Thus, the heat load calculations need to take into account the total energy needed for all of these factors. In fact, the heating system for the oil needs to be designed in excess of the theoretical load heat load calculations. Gupta et al. (2004) provided a list of items that need to be considered when estimating the heat load requirements.

- Sensible heat for heating the product entering the fryer from an initial temperature to the fryer temperature
- Moisture loss in the product from frying
- Rate of moisture loss (dehydration)
- Energy required to heat the physical systems, e.g., frying vat, piping, fittings, filtration equipment, heat exchanger, etc.
- Energy loss from radiation from the fryer and all the auxiliary equipment
- Energy loss from the fryer exhaust system
- Energy loss from the heating systems flue gases
- Thermal efficiency of the heat exchanger

The heat load requirements are thus calculated for a specific product type, production volume, and fryer system (the system includes the heating system, exhaust system, heat exchanger, pipings, fittings, etc.). An example of calculating just the frying heat load for battered and breaded chicken tender is as follows: We assume that the production rate is 5000 lb/h; thermal energy requirements are 1000 BTU/lb; heat load for frying is then – 5000 × 1000 = 5 million BTU/h. The total heat load requirements will be much greater than this number when the energy losses from the fryer and the ancillary systems are taken into consideration. In addition, a better estimate of heat load design also must include the following factors (Gupta et al., 2004):

- Delta-T, which is the temperature differential across the fryer length
- Recovery time for heating the system to maintain desired fryer temperature
- Minimization of temperature overshoot

3.4.3   Delta-T (ΔT) in the Fryer

As mentioned earlier, this is the temperature differential of the oil in the fryer between the product feed end and the product exit point. This parameter (ΔT) is a function of the loading rate and the heating capacity of the fryer. ΔT is zero when the fryer is idle and no frying is in process; as soon as the frying resumes, ΔT starts to increase and reaches a certain maximum value. Knowledge of this parameter is very important as it affects the product quality significantly. The temperature differential can be reduced by proper design of the equipment such as providing multi-zone heating in the fryer bed. In the case of indirect heating, the heated oil can be made to enter the system at more than one point in the fryer. It is not necessary to have a zero ΔT but it is important to control it and have an understanding of its effect on the product quality.

3.4.4   Conveyors

Proper selection of a conveyor system is extremely important. Different food materials require different conveyor types. Because some coated products are highly fragile, conveyor belts where there is little mechanical movement are required to avoid coating damage. Additionally, some products with delicate, tempura coatings require special conveyors known as the "tempura conveyor" or the "Teflon in-feed conveyor" (Padilla, 1998). Doughnuts, potato chips, and similar products require screen conveyors that are programmed to rotate so that both sides of the product are fried for the appropriate length of time. The conveyor length is dictated by the overall design parameters as discussed in various sections of this chapter.

3.4.5   Temperature Recovery Time

Temperature recovery time is also referred to as temperature response time of the frying system. It is the time needed to bring the oil temperature to the desired value when there is a drop in temperature from introducing the frying load. It is highly desirable to have a minimum recovery time to achieve consistent quality of the fried product. Three primary factors that influence the response time are: (1) fryer loading rate — the increased loading rate will increase ΔT, which in turn will increase the recovery time; (2) excess heating capacity of the heating system — this will reduce the recovery time depending upon the degree of excess heating capacity; and (3) design of the heating system (indirect or direct heating, single zone heating or multi-zone heating, and the thermal efficiency of the heat exchanger if employed). A temperature swing of 10 to 15°F is common in direct heating systems because the limiting factor is the heat exchange between the burner heating tubes and the oil surrounding the heating tubes. In indirect heating systems, the temperature swing is usually around 7°F, which is better than that in the direct heating systems. The least temperature difference, ±2°F, occurs in the heating systems that employ external heat exchangers. It is possible to reduce this temperature swing further by precision control of the oil flow.

3.4.6 OIL TURNOVER TIME

This parameter is expressed in hours and is the ratio of the weight of the oil in the system to the oil picked up by the product during frying. Oil turnover time is one of the prime design parameters for a frying system.

The following calculations illustrate turnover time for a system.

Total oil content in the system	= 8000 lb
Production throughput	= 4000 lb/h
Increase in the product's fat content from frying	= 20%
Oil absorbed by the product	= 800 lb/h
Turnover time	= 8000/800 = 10 h

It is desirable to minimize the turnover time because higher turnover times result in greater thermal oil degradation, which compromises the end-product quality. The time estimated above is theoretical. Thus, when factors such as the down time for maintenance or repairs, the time needed to initially heat the oil to bring it to the desired frying temperature (fryer start up time), and the time to cool the hot oil at the end of the frying period (shutdown time) are incorporated, the actual turnover time will be higher. Increase in the turnover time by 20% will proportionately increase the fryer utilization efficiency.

3.4.7 FRYER OIL FILTRATION

The importance of removing the debris and fines from the frying medium (oil) resulting from frying has already been discussed. If not continuous, a periodic removal must be done to maintain a certain level of oil quality in order to minimize the darkening of the oil and prevent burnt/undesirable flavors in the fried product. It is common to employ a continuous filtration system; in such a process, it is recommended that 5% of the oil be removed from the oil stream, passed through the filtration system, and returned to the fryer. Some systems use an active filtration approach, which is illustrated in Figure 3.10.

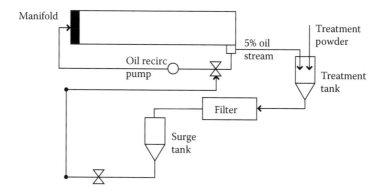

FIGURE 3.10 Schematics of an active filtration system. (Adapted from Gupta, M.K., Warner, K., and White, P.J. 2004. *Frying technology and practices*, Urbana, IL: AOCS Press.)

3.4.8 EMISSIONS

During frying, there is production of large amounts of vapors along with other volatiles emitting from cooking of the product. Installation of scrubbers is generally required as per the regulatory codes of the local, state, and federal government. Therefore, to control emissions from the fryer, the scrubbers are designed as part of the overall frying system.

3.4.9 CLEANING AND MAINTENANCE

The cleaning and maintenance protocols are integral parts of the frying operation. These protocols include sanitization, which must be done at periodic intervals. Without proper sanitization, the frying system can contain a residual amount of degraded oil, which acts as a catalyst in accelerating the oil degradation in subsequent frying. Easy cleaning and maintenance of the equipment including a cleaning in place (CIP) system is essential during frying system design and fryer selection.

3.5 IDEAL FRYER

There are specific characteristics of fryers as well as certain operating conditions that are ideal for producing the highest quality deep-fat fried foods possible. One important characteristic of the ideal fryer is size. The size of a deep-fat fryer is designated by the number of pounds of oil that is required to fill it to the frying line. An over- or undersized fryer can lead to problems with oil degradation and product quality. If the fry load is too large for the fryer, there is increased oil absorption by the food material, which leads to final products that are less healthy. Product characteristics and desired production rates must be established before choosing the appropriate fryer size.

Fryer recovery time is another key parameter of fryers. According to Durocher (1992), recovery time and closely monitored temperature control are at the heart of a good piece of frying equipment. Recovery time is the time required for a fryer to reach the set cooking temperature once a fry load has been removed. The faster the recovery time, the shorter the cooking time. Foods exposed to the oil for shorter lengths of time have less oil absorption. The rate of recovery depends on the energy requirements (BTU or kW) of the fryer. If a fryer has enough BTU or kW input to bring it back to the pre-set cooking temperature within 30 sec after a load of frozen fries is put in, then high-quality fried foods will result (Durocher, 1992). According to Durocher (1991), a good rule-of-thumb is to use a 1:6 volume ratio of food to oil in order to minimize recovery time.

The operating temperature of the fryer is the key to the frying process. The fryer must be designed so that it is able to deliver the quantity of heat necessary to cook the food. Experts recommend using a frying temperature of 350 to 360°C for frying most foods. A fryer should not overshoot the target temperature by more than a few degrees because this leads to inefficiencies in the frying process. Size, recovery time, and operating temperature of fryers all have an affect on product quality. However, fryers should also be safe, energy efficient, and easy to clean, and have an excellent filtration system in place.

3.6   RECENT DEVELOPMENTS IN FRYER TECHNOLOGY

Manufacturers are spending much time and effort trying to develop the ultimate fryer design that will deliver high-quality fried foods and extend the fry-life of the oil. The variety of deep-fat fryers that are available on the market is steadily increasing. Modern fryers are available in a number of sizes. They range from small countertop units to large, floor-mounted models. There are several fryer designs that have been introduced into the market in recent years. One recent entry has been a fully integrated floor-model fryer. This fryer does not have to be placed under an exhaust hood because it has its own built-in filtration system that removes smoke and odors from the air.

Another trend in fryer design is the addition of cool zones. A cool zone is an area at the bottom of the fryer where temperatures are cooler. Excess food particles are allowed to collect in the cool zone where they cannot impart undesirable flavors to the food that is being fried. Cool zones are most often found in larger, floor-mounted models but can sometimes be found in smaller, electric fryers.

Gas fryer manufacturers have made considerable efforts to make their fryers more energy efficient. In recent years, infrared and ceramic burners have been used to distribute heat more evenly and extract more heat energy from the fuel (Bendall, 1998). An "air fryer" is another design that has been used. These fryers are similar to convection ovens in the way they cook, and are not really considered to be fryers at all. A single-portion fryer has been introduced that can be used for low-volume restaurants or convenience stores (Durocher, 1990).

3.6.1   MODIFICATIONS OF HEAD SPACE GAS COMPOSITION AND PRESSURE

The existing restaurant-type pressure frying equipment depends on the moisture released from the products to generate the pressure inside the fryer. As this method relies on a certain amount of moisture/steam to produce the desired pressure, the use of this equipment is limited to a situation where a huge food load (~10 lb or more) is required for each single batch. In order to overcome these difficulties, researchers at Virginia Tech designed modifications to the restaurant-type pressure fryer to incorporate external gases to pressurize the frying vat (Innawong et al., 2006). A tee (T-shaped tube) replaced the exhaust tube connected between the operating valve and the fry pot for the flow of external gases. Innawong et al. (2006) conducted experiments to study the effect of using nitrogen gas on the quality of fried chicken nuggets during pressure frying in terms of moisture retention and reduction in fat uptake as influenced by frying temperature, pressure, and source of pressure generation during deep-fat frying. The time to reach the pressure generated by using nitrogen gas was significantly less (5 s) compared to using steam (ranged from 90 to 230 s). Compared to steam-released food, frying under nitrogen gas provided similar or better quality products in terms of moisture retention, juiciness, and texture. In addition to product quality improvements, frying using nitrogen gas increased the fry-life oil almost twice of that of frying using steam pressure released from the food. Similarly, research work reported by Totani (2006), Faculty of Nutrition, Kobe-Gakuin University, found that having a continuous flow of nitrogen or a slight reduction in atmospheric pressure (~97 kPa) increased the fry-life of oil by reduction in oxidative deterioration.

Further research conducted by Ballard and Mallikarjunan (2006) on breaded chicken nuggets found similar results except that frying load significantly affected the moisture retention and oil uptake in foods fried using nitrogen gas. Bengtson (2006) used the modified fryer to incorporate pressurized air or vacuum for frying breaded fish fillets and found no significant difference between nitrogen, steam, or air as a pressurizing medium in product quality attributes. However, vacuum frying resulted in juicier products compared to other frying methods.

3.7   FUTURE OUTLOOK

Moreira et al. (1999) listed some important areas that will improve frying system designs and, thus, product quality in the future. They are as follows:

- *Combination Systems:* Oven and fryer combination systems that will improve product quality by reducing oil content.
- *Enhanced System Efficiency:* Understanding how oil quality can affect heat capacity and heat transfer ability will improve fryer design, thus making systems more efficient.
- *Reduced Fat Content:* Understanding the interaction between oil chemistry and heat, mass, and momentum transfer during frying will enhance frying system designs and improve the quality of products by reducing oil content. In addition, with the FDA approval of Olestra, a zero-calorie fat replacer,

new developments will be required to understand the frying characteristics of this product.

- *Odor Control:* Means of reducing emissions from frying systems will be of great concern in the future. In Europe, for example, frying systems are allowed to emit only water and CO_2. Some examples of emissions control systems include catalytic converters, scrubbers, incinerators, and direct-fired heating of frying oil with incineration.
- *Process Control:* Automatic frying systems run with less supervision and greater productivity than systems with manual controls.
- *Online Sensors:* It is very important today to have sensors capable of measuring online in real time the quality of the frying oil (viscosity, TPM [total polar materials], FFA [free fatty acids], etc.) and frying products (oil content, crispness, color, etc.) to better control frying systems efficiently. Through feedback loops to computer-based controllers, operation conditions can be altered continuously online to ensure high-quality products.

3.8   REFERENCES

Ballard, T. and Mallikarjunan, P. 2006. The effect of edible film coatings and pressure frying using nitrogen gas on the crispness of breaded fried chicken nuggets. *J. Food Sci.* 71(3): S259–S264.

Bendall, D. 1998. Well-equipped frying equipment. *Food Management* 33(3): 72, 74–76, 78.

Bengtson, R. 2006. The effect of novel frying methods on quality of breaded fried foods. M.S. Thesis, Virginia Polytechnic Institute and State University, Blacksburg, VA.

Cummings, G. 1983. The facts of frying. *Restaurant Business* 82(6): 246, 248, 250.

Durocher, J.F. 1990. Fry stations. *Restaurant Business* 89(2): 158-160.

Durocher, J.F. 1991. In short order. *Restaurant Business* 90(14): 270, 272.

Durocher, J.F. 1992. Competitive edge. *Restaurant Business* 91(18): 158–159.

Fisher, D. 2002. *Commercial Cooking Appliance Technology Assessment*, Report No. 5011.02.26, Food Service Technology Center, Pacific Gas and Electric Co., San Francisco, CA.

Gupta, M.K., Warner, K., and White, P.J. 2004. *Frying technology and practices*, Urbana, IL: AOCS Press.

Innawong, B., Mallikarjunan, P., Marcy, J., and Cundiff, J. 2006. Pressure conditions and quality of chicken nuggets fried under gaseous nitrogen atmosphere. *J. Food Proc. Preserv.* 30: 231–245.

Jacobson, G.A. 1991. Quality control in deep-fat frying operations. *Food Technol.* 45(2): 72–74.

Mallikarjunan, P., Chinnan, M.S., and Balasubramaniam, V.M. 1997. Mass transfer in edible film coated chicken nuggets: Influence of frying temperature and pressure. In *Advances in food engineering,* Narsimhan, G., Okos, M.R., and Lombardo, S. (Eds.), West Lafayette, IN: Purdue University, pp. 107–111.

Moreira, R.G., Castell-Perez, M.E., and Barrufet, M.A. 1999. *Deep-fat frying: Fundamentals and applications.* Gaithersburg, MD: Aspen Publishers.

Padilla, J. 1998. Fryer systems technology: Direct and indirect heated systems. *Cereal Foods World* 43(8): 635–637.

Rao, V.N.M. and Delaney, A.M. 1995. An engineering perspective on deep-fat frying of breaded chicken pieces. *Food Technol.* 49(4): 138–141.

Steir, R.F. 1996. Understanding high-volume frying, Part 1. *Baking and Snack,* 18(2): 70–76.

Totani, N. 2006. A small reduction in atmospheric oxygen decreases thermal deterioration of oil during frying, *J. Oleo Sci.* 55(3): 135–141.

4 Frying Oil

Fats and oils play important functional and sensory roles in fried food products. They are responsible for carrying, enhancing, and releasing flavor of other ingredients to develop texture and mouth feel characteristics (Moreira, Castell-Perez, and Barrufet, 1999). The frying oil is the single most important factor in determining the final quality of deep-fat fried foods. In the frying process, the oil serves as the heating medium and typically reaches temperatures anywhere from 180°C to greater than 200°C. The elevated temperatures and repeated use of the frying oil increases its susceptibility to thermal and oxidative degradation. More specifically, chemical reactions such as hydrolysis, polymerization, oxidation, and fission occur. Because of the various chemical reactions, decomposition products such as free fatty acids (FFAs), hydroperoxides, surfactants, hydrocarbons, and a host of others are formed. In fact, more than 400 different chemical compounds, including 200 volatile products, have been identified in deteriorating frying oil (Moreira et al., 1999).

Decomposition products along with frying oil temperature, turnover rate, type of food material, the presence of oxygen and water, and the design and maintenance of the frying equipment are all factors that affect the rate of degradation of the frying oil. Chemical and physical changes in the oil can prolong frying time, increase oil absorption by the food material, and lower the final nutritive value of the product.

The key elements of good frying oil are a bland flavor, pale color, and good oxidative and thermal stability during the frying operation (Baskou and Elmadfa, 1999). Determining the point where these elements no longer exist is a difficult task. There have been many methods developed to determine when frying oils have deteriorated to the point where flavor, texture, and nutritional value of the food have been severely diminished, and in some cases, is no longer safe. However, no one satisfactory method for determining oil quality has been developed thus far. Sensory evaluation remains the most often used method in determining frying oil quality. The right decision of when to discard frying oil would minimize costs and deleterious effects on the quality of fried products and health (Al-Kahtani, 1991). Different countries have established different regulations about when to discard used frying oil (Table 4.1).

4.1 FRYING OIL QUALITY

Frying oil is used repeatedly at elevated temperatures in the presence of oxygen and moisture. This results in the accumulation of decomposition products that affect not only food quality, but human health as well. The term *decomposition products* refers to all degradation products with molecular weights higher than that of triglycerides. A schematic representation of oil degradation during deep-fat frying is shown in Figure 4.1. Oil is exposed to the action of four agents that cause drastic changes in its structure: (1) moisture (as steam) from food, giving rise to oxidative

Table 4.1 Regulations for Discarding Frying Medium in Several Countries

Country	Regulated	Federal/Local	Polars, %	OFA, %	FFA, %	Smoke point, C	Max. Temp, C	DPTG, %	Viscosity at 50 C, mPa.s	Appearance	Sensory attributes
Austria	Yes	Federal	27	1.0	1.3	170	180			Dark color or foaming	Unpleasant odor
Belgium	Yes	Federal	25		2.5		180	25	37		
Chile	Yes	Federal	24	1.0	1.0	170				Smokes or foaming	Unpleasant odor
Canada	No										
France	Yes	Federal	25				180				
Germany	Yes	Local	24	0.7		170					
Hungary	No										
Spain	Yes	Federal	25								
Israel	No										
Italy	Yes	Federal	25		2.5	170	180				
Japan	No										
Netherlands	Yes	Federal			2.3			16			
Portugal	Yes	Federal	25				180				
Denmark	No										
Norway	No										
Finland	No				1.3	170					
Sweden	No										
Switzerland	Yes	Local	27			170				Color	Sensory can result in rejection
USA	No				2.0					Smoking, color	

Note: Polars = Total polar compounds; OFA = Oxidized fatty acids; FFA = Free fatty acids; DPTG = Dimer and polymeric triglycerides.

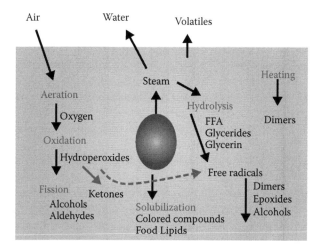

FIGURE 4.1   Oil degradation kinetics.

alteration; (2) atmospheric oxygen (present in air) entering oil from the surface of the container, giving rise to oxidative alteration; (3) high temperatures at which the frying operation takes place, which results in thermal alteration; and (4) contamination from food ingredients (Moreira et al., 1999). It has been shown that the longer the frying time and the higher the frying temperature, the higher the amount of total polar components (TPC) that will be present in the frying oil. However, if the frying temperature is too low, the food must remain in the oil for a longer time in order to fully brown. The extended time in the oil increases oil absorption by the food material. Factors affecting how heated oils and their degradation products interact with food include the food, oil, water–oil interaction, surfactants, and oxygen. If there are no changes in processing equipment and no changes in the type or volume of oil in the fryer, then all changes observed in the finished products are due to changes in frying oil (Blumenthal, 1991).

Severely oxidized oils may also affect the nutritional value of fried foods. According to Moreira et al. (1999), the nutritional value of oils is affected by the loss of polyunsaturated fatty acids, which supplement the essential fatty acid requirement in the human metabolism. Furthermore, many of the decomposition products are harmful to human health because they destroy vitamins, inhibit enzymes, and can be potential carcinogens.

4.1.1   Oil Degradation

Blumenthal (1991) researched the quality of French fries fried in oil of varying qualities. He postulated that the oil goes through five phases leading up to degradation. The phases are as follows:

1. Break-in-oil: White product; raw; ungelatinized starch at center of the fry; no cooked odors; no crisping of the surface; little oil pickup by food.

2. Fresh oil: Slight browning at the edges of the fry; partially cooked (gelatinized) centers; crisping of the surface; slightly more oil absorption.
3. Optimum oil: Golden-brown color; crisp, rigid surfaces; delicious potato and oil odors; fully cooked centers (rigid, ringing gel); optimal oil absorption.
4. Degrading oil: Darkened and/or spotty surfaces; excess oil pickup; product moving toward limpness; case-hardened surfaces.
5. Runaway oil: Dark, case-hardened surfaces; excessively oily product; surfaces collapsing inward; centers not fully cooked; off-odor and -flavors (burned).

The challenge lies in being able to maintain the oil in its optimum state for as long as possible. Attempts have been made to prolong the optimum stage by continuously adding fresh oil to used oil during the frying process. This type of method is effective up until a certain point at which any further addition of fresh oil will have no affect on the overall quality of the oil. The rate at which the oil reaches this point is referred to as the oil turnover rate.

4.1.2 CHEMICAL CHANGES IN OIL

Oil undergoes a series of chemical changes throughout the frying process. Many of these changes occur because of the presence of moisture being released from the food material and atmospheric oxygen. Atmospheric oxygen reacts with the oil at the surface, thus creating oxidative reactions. Oxidation produces hydroperoxides, which can further undergo three major types of degradation: (1) fission, which produces alcohols, aldehydes, acids, and hydrocarbons; (2) dehydration, which produces ketones; and (3) free-radical formation, which produces oxidized monomers, oxidative dimmers, and polymers, trimers, epoxides, alcohols, hydrocarbons, and nonpolar dimmers and polymers (Moreira et al., 1999). The frying oil also undergoes other chemical reactions such as hydrolysis and polymerization.

4.1.2.1 Hydrolysis

Hydrolysis is a chemical reaction that results from the interaction of triglycerides in the oil with water that is released from the food as steam. The interaction between triglycerides and steam forms low weight molecular compounds such as FFAs. The rate of FFA formation depends partly on the amount of water in the food, temperature of the oil (higher temperatures created increased FFA production), rate turnover of oil (higher turnover rate, slower production of FFAs), and the accumulation of debris and burnt food particles in the fryer. The debris tends to accelerate the development of FFAs (Boskou and Elmadfa, 1999). Other decomposition products that are produced because of hydrolysis include glycerol and mono- and diglycerides.

4.1.2.2 Polymerization

Polymerization is a chemical process that results in the formation of higher molecular weight compounds called polymers. Polymers and dimers represent degradation products that are unique to fried foods and are excellent chemical markers of oil degradation (Boskou and Elmadfa, 1999). They are nonvolatile decomposition products that are responsible for the physical changes in oil such as increased viscosity, darkened color, and foaming. The formation of polymers depends on temperature, surface-to-volume ratio of oil, heating time, and fatty acid composition of the oil (Christie et al., 1998).

4.1.2.3 Surfactants

Interfacial tension (IFT) measures the degree of interaction between the frying oil and the food material. The ability to control the level of IFT can greatly enhance the quality of the frying oil. Chemical compounds known as surfactants make it extremely difficult to control IFT. Surfactants are chemical compounds such as soaps, phospholipids, inorganic salts, and polymers that are formed from oxidative reactions in the oil. They act to increase the interaction between oil and water. In a study conducted by Gil and Handel (1995), it was found that the surfactant sodium oleate was effective in reducing the IFT in soybean oil used to fry donuts. In fact, sodium oleate reduced IFT to nearly zero at a concentration of just 0.1%. In that same study, soaps were found to have the greatest effect in reducing IFT. Blumenthal (1991) developed a surfactant theory that suggests that controlling surfactant formation is the key to maintaining the quality of the frying oil. Blumenthal's surfactant theory is based on the following assumptions:

1. Frying is basically a dehydration process. When food is fried, water and materials suspended or dissolved in the water are heated and "pumped" from the food to the frying oil.
2. The heat-transfer medium, the frying oil, is a nonaqueous material, whereas the food can be assumed almost water. Water and oil are immiscible.
3. For frying to occur, heat must be transferred from the nonaqueous medium — the oil — to the mostly aqueous medium — the food.
4. Any changes in the frying or heat transfer characteristics of the oil must result from degradation products formed from the oil.
5. Food materials leaching into the oil, thermal and hydrolytic breakdown of the oil, and oxygen absorption at the oil–water interface all contribute to altering the oil from a medium that is almost pure triglyceride to a mixture of hundreds of compounds.
6. Substances that affect heat transfer at the oil–food surface reduce the surface tension between the two immiscible materials (water and oil). These substances act as wetting agents and are regarded as surfactants.
7. As the oil degrades, more surfactants are formed, causing increased contact between the food and oil. This causes excessive oil uptake by the food and an increased rate of heat transfer to the surface of the food. Eventually, excessive darkening and drying of the surface occur before the food is cooked, as the rate of conduction of heat to the interior of the food is constant and cannot be sped up by changes in the oil.

From the surfactant theory, it can be seen that surfactants act as catalysts that enhance many of the breakdown reactions in oil (Gil and Handel, 1995).

4.1.3   PHYSICAL CHANGES IN OIL

The frying oil goes through several physical changes during the frying process. These changes include an increase in viscosity, change of color, foaming, and a reduction in the smoke point. Many of these changes occur because of polymerization, in which nonvolatile, high-molecular-weight compounds are formed. The physical changes that occur are not desirable. When the oil thickens, the rate of heat transfer is reduced. Therefore, it takes longer to cook the food and to produce the desirable golden-brown color. Frying results in darkening of the oil due to oxidation and colored pigments from the food diffusing into the oil (Al-Kahtani, 1991). The smoke point is the temperature at which oil begins to smoke continuously (Moreira et al., 1999). Smoking of oil is a direct result of the breakdown of the triglycerides in oil to form FFAs and glycerols. The higher the smoke point, the more suitable the oil is for frying.

4.2   MEASURING OIL QUALITY

Various criteria are used to judge when the frying oil needs to be discarded. In restaurants and food services, changes in physical properties of frying oils have been used as an indicator of oil quality (Moreira et al., 1999). For example, the frying oil may be discarded when the oil becomes dark, when there is too much smoke, strong odor, greased texture, or when a persistent foam layer of specified thickness is observed (Moreira et al., 1999). However, before an operator sees such effects, the oil has usually suffered considerable decomposition. In the food industry, not only physical tests but also chemical tests are used to measure oil quality including acidity, polymer content, and total polar content (Moreira et al., 1999; White, 1991). Table 4.2 summarizes various oil quality testing methods and corresponding manufacturers of instruments to perform these tests.

There has been no single method developed that can accurately predict the extent of oil degradation. A combination of methods must be used to determine when oil should be discarded. For example, tests for peroxide values and FFA content alone

Table 4.2   Rapid Test Methods to Measure Oil Quality

Device	Manufacturer and contact information
Viscosity	
FRI-CHECK	Fri-Check, Belgium (www.fri-check.de, now owned by Mir-Oil)
Oil Color	
Lovibond	Lovibond Tintometers (www.tintometer.com)
Total Polar Compounds	
PCT120	3M Corp. (www.3m.com)
Food Oil Sensor	Northern Technologies (www.ntic.com) — Currently not supported
Oxifrit and Fritest	Merck (www.merck-chemicals.com)
OptiFry	Mir-Oil (www.miroil.com)
Testo 265	Testo Ltd (www.testo.co.uk)
Food Oil Meter	Ebro Electronics (www.ebro.de)

are not sufficient. These compounds are primary oxidation products that are rapidly decomposed during the frying process. Because these compounds are not allowed to accumulate, a quantitative measurement of the presence of these compounds should not be the sole basis for deciding whether to discard oil. FFAs and hydroperoxides are further broken down into secondary decomposition products. The secondary decomposition products accumulate in the oil and, thus, better measurements of oil quality can be obtained by evaluating the amount of secondary products present in the frying oil. However, sensory evaluation remains to be the most often used method to determine the quality of frying oils.

A number of different testing methods have been developed over the years. The majority of the tests involve either chemical analysis techniques or simple, quick tests that can be carried out in a matter of minutes. The most commonly used methods include column and gas chromatography, FFA and peroxide value determination, food oil sensor, and a variety of tests that measure color such as the Fritest, Spot Test, and the RAU test.

4.2.1 CHROMATOGRAPHY

Column chromatography is often used to measure the amount of total polar materials (TPM) present in oil. In this method, a weighed amount of fat (2.5 g) is dissolved in light petroleum ether:diethyl ether (87:13) and is run through a silica gel column that absorbs the polar compounds. After evaporation of the eluted solvent, nonpolar fat is weighed and TPM is estimated by the difference. Kazemi, Wang, Ngadi, and Prasher (2005) used column chromatography according to Method Cd 20-91 (AOCS, 1989) to determine the TPCs in heated mixtures of hydrogenated and non-hydrogenated canola oils. Xu (2003) used this method on canola oil used to fry French fries and Miyagi, Nakajima, Nabetini, and Subramanian (2001) used it for membrane-processed oils. Matthaus (2006) used Method C-III 3b (DGF, 1998) to determine the TPCs in four different oils used to fry potatoes for 72 h. Benedito, Mulet, Velasco, and Dobarganes (2002) used the IUPAC Standard Method 2.507 (IUPAC, 1992) to determine TPC of heated virgin olive oil. Shyu, Hau, and Hwang (1998) separated polar and nonpolar compounds in three oils by column chromatography according to a method by Waltking and Wessels (1981), in which silica gel (70 to 230 mesh) was used as the column-packing material and the polar compounds were eluted with a solvent mixture of 87% light petroleum and 13% diethyl ether.

Gas chromatography (GC) is widely used to analyze fatty acid profiles. The method involves completely converting the oil into methyl esters. Gas liquid chromatography (GLC) analysis is then conducted. The procedure can be completed in less than one hour. Many chromatographic columns specific for analyzing oils can be purchased from vendors like Waters, Hewlett Packard, and Supelco. Fregapane, Lavelli, Leon, Kapuralin, and Salvador (2001) determined the FFA content of olive oil using European Regulation EEC 2568/91 (EEC, 1991). Matthaus (2006) determined the FFA content of four oils used to fry potatoes for 72 h by Method C-V 2 (DGF, 1998). Kita and Lisinska (2005) used GC to determine FFA of methylated frying oils according to AOAC (1995). No specific method was given; however, they used a capillary column RTX-2330 (150 m × 0.25 mm i.d., 0.20-μm film thickness). Besbes et al. (2005)

determined FFA of heated date seed oils using GC analysis with an HP-5890 Series II instrument equipped with a flame-ionization detector (Hewlett-Packard, North Hollywood, CA) and an HP Inovax capillary column (30 m × 0.32 mm × 0.25 μm film). Che Man, Ammawath, Rahman, and Yusof (2003) determined FFA by performing GLC after transesterification of palm oil used to fry banana chips. They used the same instrument as Besbes et al. (2005), HP model 3392A integrator, and the same column as Kita and Lisinska (2005), but 30 m in length. Shyu et al. (1998) used capillary GC analysis with the same instrument and column as Besbes et al. (2005), but with an inner diameter of 0.25 mm and film thickness of 0.5 μm, to determine FFA composition of methylated oil samples according to a method described by Christie (1982). Innawong, Mallikarjunan, Irudayaraj, and Marcy (2003) used alkaline titration from Method Ca 5a-40 (AOCS, 1989) to determine FFA of peanut oil from a fast food restaurant. Holownia, Chinnan, Erickson, and Mallikarjunan (2000) used the same method to determine FFA of peanut oil before and after frying chicken strips coated with edible films, and Innawong (2001) used it to determine the effect of pressure conditions on commercial vegetable oil quality.

High performance liquid chromatography (HPLC) has been used to characterize polymeric compounds in frying oils. Matthaus (2006) determined the oligomer triacylglycerols of four oils used to fry potatoes for 72 h by HPLC according to Method Cd 22-91 (AOCS, 1989). Gertz (2004) used high performance exclusion chromatography (HPSEC) from Method C-III 3c (DGF, 2002) to determine triacylglycerols, dimers, and polymers in sunflower oil containing oil additives. Benedito et al. (2002) also used HPSEC, but followed the IUPAC standard method (IUPAC, 1992). Gertz (2004) determined peroxide value (POV) of sunflower oil containing various oil additives using an iodometric assay from method C-VI 6a (DGF, 2002).

4.2.2   RAU Test

This colorimetric test kit contains redox indicators that react with the total amount of oxidized compounds in a fat sample. The color of the mixture of sample and reagent is compared to a four-color scale with 1 representing good, 2 still good, 3 intermediate quality, and 4 bad.

4.2.3   Fritest and Oxifrit

Both of these tests were developed by Merck to provide fryer operators and regulators a quick means to monitor oil quality The Fritest measures the alkali color number and the Oxifrit measures the amount of oxidized products. Both tests are colorimetric tests using solvent-based reagents and are sensitive to carbonyl compounds in the oil. The mixture of the sample and reagent is compared to diagnostic colors with Fritest using a three-color scale and Oxifrit using a four-color scale. In Fritest, the test reagent and oil sample are added to predetermined levels and mixed before reading the color against the scale. For the Oxifrit test, five drops of another reagent are added to the sample holder before adding the oil. These tests are currently being used by regulators in Switzerland, Finland, Denmark, Austria, Luxembourg, Portugal, Norway, and Sweden.

4.2.4 COLOR

Oil color can be measured with a Lovibond Tintometer Model PFX990 (HF Scientific Inc., Ft. Meyers, FL) using a cell with an optical path length of 25.4 mm (1 in.), which requires 12 mL of oil sample (Figure 4.2). Results from the Tintometer are expressed in an AOCS Red, Yellow scale. The oil samples have to be heated to 45°C to assure that they are in the liquid state before measurement. Transmission of light through the sample and sample cell is affected by any change in their temperatures. Therefore, to minimize any experimental errors from such temperature variations, the sample holding cell and the Lovibond Tintometer should be maintained at 45°C with the help of a built-in temperature control mechanism. After taking readings for each sample, the cell can be reused after rinsing with hot tap water. In addition to the Lovibond tintometer, spectrophotometers and spectrophotocolorimeters have been used to describe color changes in frying oil.

Besbes et al. (2005) determined the color (L*, a*, b*) of heated date seed oils using a spectrophotocolorimeter MS/Y-2500 (Hunterlab, Inc., Reston, VA). Xu (2003) used a Minolta CR300 chromameter (Minolta Camera Co. Ltd., Osaka, Japan) to measure the color of four oils used to fry French fries. While the meter displayed L*, a*, and b* coordinates, only L* coordinates were used as these were determined to have a significant correlation to the TPC content of the oils. Miyagi et al. (2001) measured the color of membrane-processed oils using a spectrophotometer (Model V-570; Jasco Corp., Tokyo, Japan). Holownia et al. (2000) used a Lovibond Tintometer, Model PFX990 (HF Scientific, Inc., Ft. Meyers, FL), with a 1-in. (25.4 mm) optical path length from Method Cc 13e-92 British Standard (AOCS, 1989) to measure the color of peanut oil before and after frying chicken strips coated with edible films. Benedito et al. (2002) used this method for virgin olive oil and Chu and Hsu (2001) used it for shallot-flavored frying oil.

4.2.5 SPOT TEST

This is a colorimetric procedure in which a drop of oil is placed on a silica-gel-covered slide that has bromocresol green incorporated into the gel as a pH indicator. The test monitors FFA content as an indicator of hydrolytic rancidity. Diagnostic colors of the pH indicator are blue, green, and yellow.

FIGURE 4.2 Instrument to measure oil color — Lovibond tintometer. (Courtesy of Tintometer Limited, Wiltshire, UK.)

4.2.6 Food Oil Sensor

The food oil sensor has been on the market for more than 30 years. It is a compact and portable machine that is used to measure the dielectric constant in degrading oils relative to fresh oils. A sensitive bridge circuit detects the changes in the dielectric constant corresponding to the polar compounds in the used frying oil (Figure 4.3). The changes are measured in reference to fresh oil and thus a calibration with fresh oil is an important step. The dielectric constant is directly proportional to the amount of degradation products in the oil. As the method uses calibration corresponding to the type of oil being studied, the numbers cannot be easily correlated to the total polar compounds. The other concerns include the sensitivity to air currents, long warm-up period, possible adverse effects of water on results, and effects of contamination in frying oil from product fat. Northern Instruments, which manufactured the food oil sensor, has elected to discontinue making them. To fill the gap in this area, three new measurement devices using the same dielectric property as a means to monitor oil quality were introduced. The MirOil, a U.S. company, introduced OptiFry (Figure 4.4); The Ebro, a

FIGURE 4.3   Food oil sensor. (Courtesy of Northern Instruments, Lino Lakes, MN.)

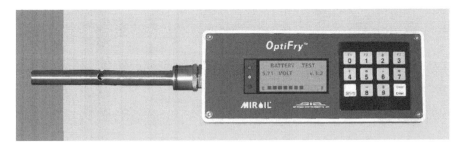

FIGURE 4.4   OptiFry. (Courtesy of Miroil, Allentown, PA.)

German company, introduced the food oil monitor (FOM) (Figure 4.5); and Testo AG, another German Company, has Testo 265. These units need to be calibrated with each type of oil in which they will be used. The user turns on the unit, inserts the probe in the hot oil to be tested, and reads the results directly on the digital display.

4.2.7   POLAR COMPOUND TESTER

The polar compound tester manufactured and introduced in 2000 by 3M Corp. allows users to evaluate the percentage of polar materials in degrading oils based on changes happening in the test strip (Figure 4.6). After placing the test strip in the metallic well,

FIGURE 4.5   Ebro food oil monitor. (Courtesy of Ebro Electronics, Ingolstadt, Germany.)

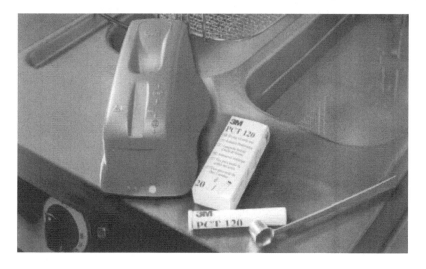

FIGURE 4.6   Polar compounds tester. (Courtesy of 3M Corporation, Minneapolis, MN.)

the user has to pour the oil through a funnel and get the result from the embossed lid in the instrument. 3M believes that the system is simple to use with precision required to satisfy the regulatory requirements in Europe for restaurant frying oils.

4.2.8   VISCOSITY

Viscosity is considered one of the most important physical properties of oils; consequently many researchers have studied it (Lang, Sokhansanj, and Sosulski, 1992; Toro-Vazquez and Infante-Guerrero, 1993; Miller, Singh, and Farkas, 1994; Paul and Mittal, 1997; Tseng, Moreira, and Sun, 1996). Miller et al. (1994) reported kinematic viscosity of oil at 170 to 190°C. They studied viscosities of frying oils to correlate them with oil quality. Toro-Vazquez and Infante-Guerrero (1993) also reported Newtonian behavior of oil. Lang et al. (1992) measured kinematic viscosity of canola oil and proposed a model for calculation of viscosities over a temperature range of 4 to 100°C as shown in the following.

$$v = \exp\,(C_0 + C_1\,T + C_2\,T^2)$$

However, the report by Lang and co-workers does not explain the changes in viscosity of canola oil above 100°C. Tseng et al. (1996) studied the viscosity of fresh and abused soybean oil at 25 to 190°C. In order to select a filter system, it is important to understand flow behavior of the fluid to be filtered at the temperatures at which the filtration is carried out. The experiment conducted by Bheemreddy, Shinnan, Pannu, and Reynolds (2002a) was intended to give the flow behavior of canola oil from room temperature to that of frying conditions. An Arrehnius equation shown in the following was fitted to the viscosity data.

$$\eta = \eta_0\,\exp(-Ea/RT)$$

Regression analysis of $y = \log\,(\eta/\eta_0)$ and $x = (1/T)$ showed an excellent fit. The regression equations are:

For fresh oil, $y = 2683.2$ $x - 4.76$ $(R^2 = 0.993)$
For used oil, $y = 2468.2$ $x - 4.28$ $(R^2 = 0.998)$

Besbes et al. (2005) determined the viscosity of heated date seed oils at 25°C using a Stress Tech Rheologica Rheometer (Rheologica Instruments AB, Lund, Sweden). Kazemi et al. (2005) used a controlled stress/shear rheometer with a coaxial concentric cylinder and built-in peltier plate (Advanced rheometer, AR 2000, TA Instruments Inc., Leatherhead, UK) to measure mixtures of hydrogenated and non-hydrogenated canola oils at 40°C. Tomasevic and Siler-Marinkovic (2003) estimated the viscosity of refined sunflower oil and used frying oils by Method DIN 5156 using a Vogel-Ossag-type viscosimeter (DIN, 1956). Benedito et al. (2002) measured the viscosity of heated virgin olive oil in the range of 25 to 35°C and Miyagi et al. (2001) measured the viscosity of membrane-processed oils at 25°C using a falling-ball viscometer (Microviscometer, HAAKE Corp.,

Germany). Holownia et al. (2000) used a Brookfield Digital Viscometer, Model LVTDV-II (Brookfield Engineering Laboratories, Inc., Stoughton, MA) with an attached cylindrical spindle #1 LV to measure peanut oil viscosity before and after frying chicken strips coated with edible films according to Method Ja 10-87 (AOCS, 1989). Shyu et al. (1998) used a Brookfield Synchro-Lectric Viscometer, Model RVF (Brookfield Engineering Laboratories, Inc., Stoughton, MA) to determine the viscosity of soybean oil (25°C), palm oil (40°C), and lard (40°C).

4.2.9   FRI-CHECK UNIT

This unit measures viscosity and is correlated to polar content in the frying oil (Figure 4.7). The unit contains an electronic box with a removable steel tube in which the oil sample has to be placed. The time taken by a falling piston type body through the oil is correlated to polar compounds. The unit takes approximately 5 to 7 min to complete the analysis and give a result. The form of the falling body is cylindrical and the diameter is almost the same as that of the tube. The thermo-oxidative changes in the frying fat during heating can be monitored continuously using Fri-Check. Gertz (2000) compared the results from Fri-Check to standard methods with more than 100 samples collected from fast food restaurants and found it to have better correlation with TPCs (0.92) than with polymerized triglycerides (0.88).

FIGURE 4.7   Fri-Check. (Courtesy of Fri-Check bvba, Hulshout, Germany.)

4.2.10   Acid Value

Kazemi et al. (2005) used alkaline titration according to Method Cd 3a-63 (AOCS, 1989) to determine the acid values of heated mixtures of hydrogenated and non-hydrogenated canola oils. This method was also used by Tomasevic and Siler-Marinkovic (2003) on refined sunflower oil and used frying oils, by Chu and Hsu (2001) on oil used to fry shallots, and by Shyu et al. (1998) on three oils (soybean, palm, and lard) used to fry carrot slices under vacuum.

4.3   RECENT DEVELOPMENTS IN IMPROVING OIL QUALITY

4.3.1   Prolonging the Frying Oil Life

Off-flavorings, nutritional losses, and other deteriorative changes in oil will increase by reaction with oxygen or by hydrolytic reactions. Cold storage, good transportation, careful packaging, and sterilization can minimize the effects of hydrolytic reactions. However, oxidative rancidity cannot be stopped by the above methods alone because it is a chemical reaction with low activation energy. Research into the problems of oxidative deterioration has been conducted for many years. Peter and Hakan (1998) reported such oxidations could cause damage to cell membranes and DNA and cause cancer growth (Navarro et al., 1999).

Auto-oxidation can be inhibited or retarded by vacuum packaging, packing under an inert gas to exclude oxygen, and refrigeration/freezing. The use of antioxidants is the most preferable way of suppressing the oxidation (Milić, Djilas, and Čanadanović-Brunet, 1998; Fuster, Lampi, Hopia, and Kamal-Eldin, 1998; Rudnik et al., 2001) because the above methods are not always applicable and it is seldom realized how little oxygen is needed to initiate and maintain the oxidative process. Naz, Sheikh, Siddiqi, and Sayeed (2004) studied oxidative stability of olive, corn, and soybean oil and determined the exact onset of oil deterioration under the conditions they were normally stored. They observed decline in the rate of oxidation following addition of plant extracts, such as tea (*Camellia sinensis*), and different purified phenolic acids (ferulic, caffeic, vanillic, and gallic acids) as antioxidants. Rosemary and sage are two plant-derived antioxidants that have been studied intensively and proven to be effective for stabilizing frying oils (Che Man & Tan, 1999). Rosemary and sage, when added to palm olein, effectively retarded oil deterioration during a 6-day deep-fat frying of potato chips. Both antioxidants increased the acceptability of the fried product (Che Man and Jaswir, 1999). Together with citric acid, rosemary and sage can exhibit a synergistic effect to retain the fatty acid composition of palm olein during repeated deep-fat frying (Jaswir, Che Man, and Kitts, 2000a). Jaswir, Che Man, and Kitts (2000b) reported that the use of oleoresin rosemary extract, sage extract, and citric acid improved the sensory scores of potato chips during a 5-day repeated deep-fat frying.

Currently, the use of filter aid is a practical and efficient method that the frying industry employs to extend the life of frying oils and improve healthy aspects of used frying oils. There is some research in using commercial synthetic adsorbents to maintain oil quality and recover used oils. Filter aid materials or their combinations were found effective for the control of FFAs and color of used frying oils (Yates & Caldwell,

1992, 1993). Lin, Akoh, and Reynolds (2001) studied the two selected combinations of four commonly used filter aids being used to recover used frying oils by vacuum filtration: Britesorb (Br), Hubersorb 600 (HB), Frypowder (Fr), and Magnesol (Ma). The effectiveness of the adsorbent treatments was evaluated by refrying chicken fritters and breasts in the recovered used oil. Such adsorbents combination may be useful to enhance the quality of fried food and prolong oil life.

Purification of used oil can be achieved by removing the undesirable oxidized materials and polymers. Some attempts have been made to purify used frying oil. Jung and Rhee (1994) and Lin, Akoh, and Reynolds (1999) purified oil by adsorbent treatments. Yoon et al. (2000) tried supercritical fluid extraction (SFE) methods to separate various types of organic compounds and purified the used oil. Major advantages of the SFE process are low temperature separation, low energy consumption, and good selectivity (Yoon et al., 2000).

4.3.2 FILTRATION AND FILTER AIDS

Scientific literature is replete with studies addressing the safety of used frying fats and oils. The fact that some of the fat is absorbed by every piece of food fried necessitates the use of a good quality frying medium and maintaining it in that state as long as possible. It is estimated that 50% (or more) of the deep frying fat used in food service operations is discarded after use (Hunter and Applewhite, 1993). The discarded fryer fat is collected and recycled to be processed for animal feed and fatty acid feed stock and for its fuel value (Boyer, 1996). Improper monitoring of oil in frying operations either puts the public health at risk or causes financial losses to the operators. Thus, it is of importance to remove fat decomposition products that are harmful to the health of humans and animals.

The fried food industry is always in demand of significant process development methods to extend the fry life of oil. In recent years, demands placed on the frying industry for better quality, along with the switch to vegetable oils, have created the need for filter adsorbents or active media (Brooks, 1991). Active media (clay, zeolite, alumina, silicates, etc.) are those filter media that aid in the removal of particles and adsorb oil-soluble degradation products (Phogat, Mittal, and Kakuda, 2006). Filtration utilizing active filter media is called active filtration. Passive media are the inert materials that aid in filtration such as metal screens, rolling paper filters, plastic cloths, diatomaceous earth, etc., which remove only the insoluble particles and have little effect on the actual chemistry of the frying oil. Filtration done by using passive media is called passive filtration.

Bheemreddy et al. (2002a) established the active filtration parameters such as temperature for filtration, dosage level, duration of treatment, and multiple filtration effect, and examined the filtration systems for routine use in food service establishments. Fresh canola oil and canola oil discarded after frying corn dogs in them were used to establish the filtration parameters to use adsorbents such as Hubesorb 600 (Huber Corp., Havre de Grace, MD), Frypowder (MirOil company, Allentown, PA), and Magnasol (Dallas Group of America, Inc., Jeffersonville, IN). Laboratory-scale vacuum filtration was used to establish the parameters for active filtration, such as dosage level, duration of treatment, and the number of filtration passes. The results were implemented into a pilot scale filtration setup by modifying a restaurant-type fryer/filtration system.

HB 600 (2%) was most effective in reducing FAA followed by 2% UGA blend and 2% Magnasol. However, in the case of improvement of red color, 2% UGA blend performed better followed by 2% Magnasol. HB 600 showed the most improvement in FOS (food oil sensor) values. Magnasol and Frypowder exhibited similar results in terms of reduction in FOS values, with 12 to 14% improvement compared to the used oil. In general, the range of FFA reduction with 2% adsorbent was 25 to 35%; Frypowder was the exception, showing only 4% improvement, which was expected due to its citric acid component (Lin, Akoh, and Reynolds, 1998). The range for FFA reduction with 1% adsorbent was 15 to 20%, except for the Frypowder. In the case of red color reduction, 2% adsorbent gave 18 to 29% improvement and 1% gave 6 to 19% improvement.

4.3.2.1   Effect of Sequential Addition of Adsorbent

The rate of oil quality improvement with sequential addition of adsorbent should give an indication of the effect of daily treatment of the oil with adsorbent. Results show improvement of oil quality parameters due to sequential addition of adsorbent. When the percent improvement values were subjected to regression analysis, all four oil quality parameters showed a linear correlation in oil quality improvement with each addition of adsorbent. Oil quality parameters decreased with increasing number of adsorbent treatment. It was also found that the active treatment of used oil with higher dosage was more effective when the level of abuse of the used oil was higher; however, if active treatment were to be used on a routine basis, it would be more convenient to use a smaller dosage.

4.3.2.2   Effect of Filter Aids on Oil Quality Abused
under Commercial Frying Conditions

Bheemreddy, Chinnan, Pannu, and Reynolds (2002b) studied the effect of daily adsorbent treatment on quality and fry life of oil under simulated food service conditions. To simulate food service conditions, chicken patties were fried in canola oil in a commercial fryer for 10 days. The fryer was kept switched on for 8 to 9 h per day. Eighteen batches of frozen chicken patties were fried at 15-min intervals. One fryer oil was treated using Hubesorb 600 (J.M. Huber Corp., Havre de Grace, MD), Frypowder (MirOil Co., Allentown, PA), or Magnasol (Dallas Group of America, Inc., Jeffersonville, IN).

At the end of each day after the last batch of frying, a gravity filtration method was used to filter oil in the control fry pot. The filter paper material from Henny Penny Corp. (Eaton, OH) was used for the control samples. The oil from the fry pot was filtered in order to remove the food crumbs and breading sediments. Similarly, oil in the treated fry pot was gravity filtered, weighed, and then treated with 1% (w/w) adsorbent. An automobile-type oil filter canister and cartridge (NAPA gold filter #1453, Gard Corp., Gastonia, NC) was used to filter the adsorbent mixed oil. The oil samples were removed from the fryer for analytical tests.

A typical response expected from these test results is given in Figure 4.8. Oil sample collection time for Day 3 and the start of Day 4 is depicted in the figure with letters B, M, and E representing beginning, middle, and end of the day, respectively. The letter X represents the sample collected after active treatment. The letters B, E, M, and X are subscripted with a number indicating the day of frying.

The results show that the FFA content of both the control and the treated fry pot increased with frying time (Figure 4.9). The final level of FFA in the active treated oil was 0.5, which is one-third the amount of the control. The final AOCS red values for treated and control were 5.2 and 10.9, respectively, which indicates a 52% reduction in red color due to active treatment (Figure 4.10).

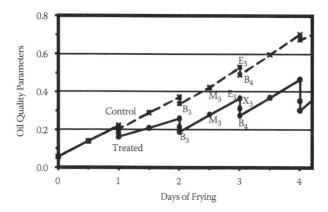

FIGURE 4.8 Graphical representation of expected changes in oil quality parameters as affected by frying time and chemical adsorbent treatment (numerical values in the graph are typical of % FFA). *Source*: Bheemreddy, R.M., Chinnan, M.S., Pannu, K.S., and Reynolds, A.E. 2002. Filtration and filter system for treated frying oil. *J. Food Proc. Eng.* 25(1): 23–40.

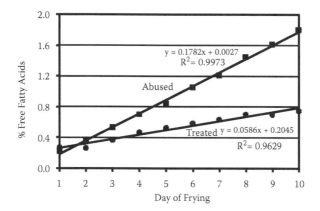

FIGURE 4.9 Change in FFA content of frying oil as affected by frying time and adsorbent treatment (vertical bars refer to standard deviations). *Source*: Bheemreddy, R.M., Chinnan, M.S., Pannu, K.S., and Reynolds, A.E. 2002. Filtration and filter system for treated frying oil. *J. Food Proc. Eng.* 25(1): 23–40.

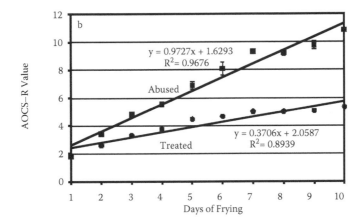

FIGURE 4.10 Change in oil color of frying oil as affected by frying time and adsorbent treatment. Red (AOCS-R) color values (vertical bars refer to standard deviations). *Source*: Bheemreddy, R.M., Chinnan, M.S., Pannu, K.S., and Reynolds, A.E. 2002. Filtration and filter system for treated frying oil. *J. Food Proc. Eng.* 25(1): 23–40.

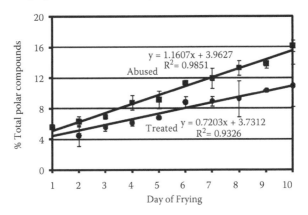

FIGURE 4.11 Change in TPC content of frying oil as affected by frying time and adsorbent treatment (vertical bars refer to standard deviations). *Source*: Bheemreddy, R.M., Chinnan, M.S., Pannu, K.S., and Reynolds, A.E. 2002. Filtration and filter system for treated frying oil. *J. Food Proc. Eng.* 25(1): 23–40.

As expected, the TPC contents of both abused and treated oil increased with frying time (Figure 4.11). The final level of TPC at the end of 10 days of frying was 16.2 % and 11.4 %, respectively, for control and treated oil samples. The final level of TPC in the treated oil was 30% less than the amount in the control oil sample.

Daily treatment of frying oil with adsorbent can extend the frying life of the oil. There was 30% less accumulation of TPC, 52% reduction in development of color (AOCS, R), and 72% lower content of FFA in frying oil due to daily active treatment of the oil.

4.4 RECENT DEVELOPMENTS IN OIL QUALITY MEASUREMENT

4.4.1 Electronic Nose

Previous monitoring methods used to analyze the volatile compounds and aroma in food needed either a highly trained sensory panel or gas chromatography/mass spectrometry (GC/MS) techniques (Hodgins and Simmonds, 1995; Hodgins, 1996; Mielle and Marquis, 1998; Yang, Han, and Noh, 2000). Each technique has its limitations. The sensory evaluation is time consuming and expensive due to dependence on the training of the judges and the description method used. The results from GC/MS still are difficult to match and directly relate to the quality of the food (Hodgins, 1996; Yang et al., 2000). Thus, there has been a genuine need for a quick, simple, and powerful objective test for indicating deterioration of oil.

Interest exists in using an electronic nose to investigate odor and volatile compounds in edible oil (Guadarrama et al., 2001; Aparicio, Rocha, Delgadillo, and Morales, 2000; Muhl, Demisch, Becker, and Kohl, 2000; Yang et al., 2000). Aparicio et al. (2000) used an electronic nose system based on conducting polymers to detect the rancidity in virgin olive oil. They found that the electronic nose provided a very fine discrimination when compared to the results from a trained panel test. Yang et al. (2000) used a commercially available portable system having six metal oxide sensors to detect rancidity in soybean oil. Muhl et al. (2000) used three gas sensors in their system to study the deterioration of frying fats. Their work compared fresh and used oil; however, the degree of usage of the frying oil was not mentioned in the paper.

Innawong, Mallikarjunan, and Marcy (2003) studied the use of an electronic nose based on quartz microbalance sensors (Figure 4.12) to differentiate oil on varying degrees of usage level in the frying oil (good, marginal, and discarded oil) instead of

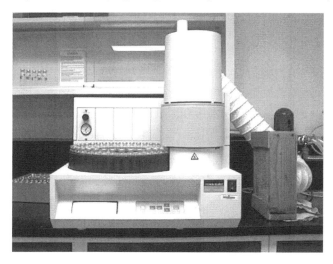

FIGURE 4.12 Quartz microbalance-based (qmb) electronic nose system for oil quality evaluation. (Courtesy of Food Engineering Lab, Virginia Tech, Blacksburg, VA.)

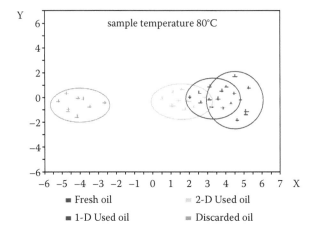

FIGURE 4.13 Canonical discrimination of frying oil by a quartz crystal microbalance sensor-based electronic nose system. *Source*: Innawong, B., Mallikarjunan, P., and Marcy, J.E. 2003. The determination of frying oil quality using a chemosensory system. *Lebensm-Wiss. u-Technol.* 37: 35–41.

just comparing fresh and used oils. The best discrimination among the three types of oils was found at a sample temperature of 80°C and a sensor temperature of 25°C. The fresh oil had a clear separation from the rest of the used oil samples (Figure 4.13). In addition, a surface acoustic wave (SAW) sensor based electronic nose was used to monitor storage stability of palm olein (Gana et al., 2005). They observed high correlation between electronic nose response and chemical and sensory evaluation data. Based on these literature data, it should be clear that the electronic nose system can be an effective method to discriminate the frying oil quality and can be used to determine whether the oil should be discarded.

4.4.2 FOURIER TRANSFORMATION INFRA RED (FTIR) SPECTROSCOPY

Recently, Fourier transformation infrared (FTIR) spectroscopy has gained widespread use for characterizing oils. This chemometric method relates to the chemical compounds in oil very well through specific peaks at specific wavelengths and analysis of the complete signal through advanced statistical packages (Figure 4.14). FTIR spectra in the mid-infrared region consist of fundamental and characteristic bands whose frequencies and intensities can clearly determine the relevant functional groups in frying oil samples. The spectra obtained from FTIR have strong C-H absorption between 3000 and 2850 cm^{-1}. Figure 4.15 clearly shows separate bands that correspond to asymmetrical C-H stretching (CH$_2$) at 2929 cm^{-1} and symmetrical C-H stretching (CH$_2$) at 2856 cm^{-1}, with a weak shoulder at 2954 cm^{-1} caused by the methylene asymmetrical stretching band. Frying oil samples also showed strong bands at 1749, 1464, and 1165 cm^{-1} that correspond to C=O (ester) stretching, C-H bending (scissoring), and C-O, CH$_2$ stretching, bending, respectively.

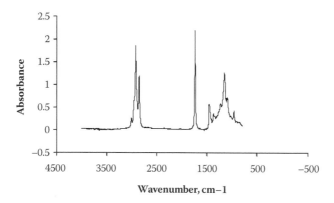

FIGURE 4.14 The FTIR-ATR spectra of frying oils in the mid-infrared range.

FTIR represents an important tool used for quality control and monitoring process in the food industry because it is less expensive, performs better, and is easier to use than other methods (Van de Voort, Sedman, and Ismail, 1993). FTIR has been used for quantitative and qualitative measurements of edible oils and fats as the results of the development of instrumental macro programming for automated routine operations and the application of chemometrical techniques for multicomponent analysis (Goburdhun, Jhaumeer-Laulloo, and Musruck, 2001; Van de Voort et al., 1994; Yang and Irudayaraj, 2000). FTIR has been used to determine *cis* and *trans* content, iodine value, saponification number, peroxide value, anisidine value, free fatty acid content of oil and fat samples, and food compositions (Dubois et al., 1996; Van de Voort et al., 1993, 1994). Limited research has been performed on used oil from food service institutions. In addition, most researchers concentrated on the particular wave numbers instead of the whole FTIR spectral data for characterizing the oil quality.

Normally, the critical absorption bands associated with common oxidation end products from the frying process could be observed in the region of 3800 to 3200 cm^{-1}, OH stretching region. The peak at 3471 cm^{-1} has been reported to be associated with the OH stretching vibration of hydroperoxide, and a weaker peak, or shoulder, in the region at 3300 cm^{-1} indicates the formation of FFA (Goburdhun et al., 2001).

Innawong et al. (2003) used FTIR to characterize frying oil samples collected from three different fast food services. In addition, they evaluated dielectric constant, peroxide value, FFAs, and density of frying oils. The correlation plots between FTIR spectra (at 3471 cm^{-1} and 3300 cm^{-1}) and chemical indexes (PV [Peroxide value] and FFA) showed the trend of higher absorbance bands at both regions (Figure 4.15) with respect to deterioration of frying oils. The correlation (r) between FTIR absorbance at 3300 cm^{-1} and FFA was from 0.84 to 0.94 for oils from the three restaurants. Similarly, the correlation between FTIR absorbance at 3471 cm^{-1} and PV was from 0.90 to 0.97 for oils from the same three restaurants. When the whole spectral data (wave numbers between 4000 and 850 cm^{-1}) of various oil rancidities were used to characterize and classify the good, marginal, and unacceptable frying oils using

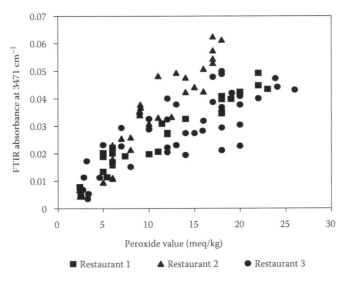

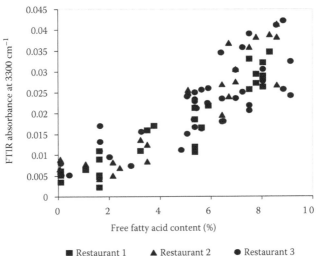

FIGURE 4.15 The correlation between chemical properties (PV and FFA) and FTIR absorbance. *Source*: Innawong, B., Mallikarjunan, P., Irudayaraj, J., and Marcy, J.E. 2003. The determination of frying oil quality using Fourier transform infrared attenuated total reflectance. *Lebensm-Wiss. u- Technol.* 37(1): 23–28.

principal component analysis, the method successfully differentiated and identified the FTIR spectra between the discarded and fresh oils (Figure 4.16).

FTIR-ATR technique would eliminate the use and disposal of hazardous solvents and reagents required by the chemical methods. The FTIR analysis is automatable and capable of analyzing approximately one sample per minute to carry out a large number of FFA analyses per day (Al-Alawi, van de Voort, and Sedman, 2004).

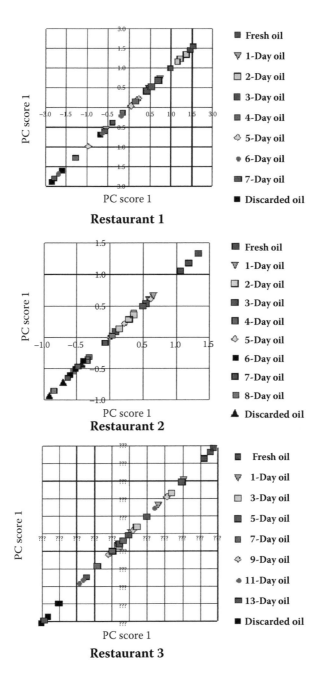

FIGURE 4.16 Principal component analysis plots for various types of oils collected from three different fast food services (between PC 1 and PC 1). *Source*: Innawong, B., Mallikarjunan, P., Irudayaraj, J., and Marcy, J.E. 2003. The determination of frying oil quality using Fourier transform infrared attenuated total reflectance. *Lebensm-Wiss. u- Technol.* 37(1): 23–28.

4.5   REFERENCES

Al-Alawi, A., van de Voort, F.R., and Sedman, J. 2004. New FTIR method for the determination of FFA in oils. *J. Am. Oil Chem. Soc.* 81(5): 441–446.

Al-Kahtani, H.A. 1991. Survey of quality of used frying oils from restaurants. *J. Am. Oil Chem. Soc.* 68(11): 857–862.

AOAC. 1995. *Official methods of analysis of AOAC international.* 16th ed., vol. 2, Arlington, VA.

AOCS. 1989. *Official methods and recommended practices of the American Oil Chemists' Society.* 4th ed., Champaign, IL.

Aparicio, R., Rocha, S.M., Delgadillo, I., and Morales, M.T. (2000). Detection of rancid defect virgin olive oil by electronic nose. *J. Agri. Food Chem.* 48(3): 855–860.

Benedito, J., Mulet, A., Velasco, J., and Dobarganes, M.C. 2002. Ultrasonic assessment of oil quality during frying. *J. Agric. Food Chem.* 50: 4531–4536.

Besbes, S., Blecker, C., Deroanne, C., Lognay, G., Drira, N-E., and Attia, H. 2005. Heating effects on some quality characteristics of date seed oil. *Food Chem.* 91: 469–476.

Bheemreddy, R.M., Chinnan, M.S., Pannu, K.S., and Reynolds, A.E. 2002a. Active treatment of frying oil for enhanced fry-life. *J. Food Sci.* 67(4): 1478–1484.

Bheemreddy, R.M., Chinnan, M.S., Pannu, K.S., and Reynolds, A.E. 2002b. Filtration and filter system for treated frying oil. *J. Food Proc. Eng.* 25(1): 23–40.

Blumenthal, M.M. 1991. A new look at the chemistry and physics of deep-fat frying. *Food Technol.* 45(2): 68–71, 94.

Boskou, D. and Elmadfa, I. 1999. *Frying of food: Oxidation, nutrient and nonnutrient antioxidants, biologically active compounds and high temperatures.* Lancaster, PA: Technomic Publishing Company.

Boyer, M.J. 1996. Environmental concerns. In: *Deep frying: Chemistry, nutrition and practical applications*, Perkins, E.G. and Erickson, M.D. (Eds.). Champaign, IL: AOCS Press, pp. 335–342.

Brooks, D.D. 1991. Some perspectives on deep-fat frying. *INFORM* 2(12): 1091–1095.

Che Man, Y.B. and Tan, C.P. 1999. Effects of antioxidants on changes in RBD palm olein during deep-fat frying of potato crisps. *J. Am. Oil Chem. Soc.* 76(3): 331–339.

Che Man, Y.B. and Jaswir, I. 2000. Effect of rosemary and sage extracts on frying performance of refined, bleached and deodorized (RBD) palm olein during deep-fat frying. *J. Sci. Food Agric.* 69(3): 301–307.

Che Man, Y.B., Ammawath, W., Rahman, R.A., and Yusof, S. 2003. Quality characteristics of refined, bleached and deodorized palm olein and banana chips after deep-fat frying. *J. Sci. Food Agric.* 83: 395–401.

Christie, W.W. 1982. *Lipid analysis.* 2nd ed., Oxford: Pergamon Press, pp. 51–53.

Christie, W.W., Dobarganes, M.C., Lavilloniere, F., Marquez, R.G., Martin, J.C., Nour, M., and Sebedio, J.L. 1998. Effect of fatty acid positional distribution and triacylglycerol composition of lipid by-products formation during heat treatment. I. Polymer formation. *J. Am. Oil Chem. Soc.* 75(9): 1065–1071.

Chu, Y-H. and Hsu, H-F. 2001. Comparative studies of different heat treatments on quality of fried shallots and their frying oils. *Food Chem.* 75: 37–42.

Corey, M.L., Gerdes, D.L., and Grodner, R.M. 1987. Influence of frozen storage and phosphate predips on coating adhesion in breaded fish portions. *J. Food Sci.* 52(2): 297–299.

DGF. 1998. Deutsche einheitsmethoden zur untersuchung von fetten, fettprodukten, tensiden und verwandten stoffen. Stuttgart, Germany: Wissenschaftliche Verlagsgesellschaft.

DGF. 2000. Deutsche einheitsmethoden zur untersuchung von fetten, fettprodukten, tensiden und verwandten stoffen. Stuttgart, Germany: Wissenschaftliche Verlagsgesellschaft.

DGF. 2002. Deutsche einheitsmethoden zur untersuchung von fetten, fettprodukten, tensiden und verwandten stoffen. Stuttgart, Germany: Wissenschaftliche Verlagsgesellschaft.

DIN. 1956. Deutscher Normenausschuss DIN Taschenbuch 20, Mineralol-und Brennstoffnormen Beuth-Wertrieb, Berlin, Koln, Frankfurt, 1956, p. 211.

Dubois, J., Van de Voort, F.R., Sedman, J., Ismail, A.A., and Ramaswamy, H.R. 1996. Quantitative Fourier transform infrared analysis for anisidine value and aldehydes in thermally stressed oil. *J. Am. Oil Chem. Soc.* 73(6): 787–794.

EEC. 1991. European Union Commission Regulation EEC 2568/91 on the Characteristics of Olive and Olive Pomace Oils and Their Analytical Methods, *Off. J. Eur. Comm.* L248.

Fregapane, G., Lavelli, V., Leon, S., Kapuralin, J., and Salvador, M.D. 2001. Effect of filtration on virgin olive oil stability during storage. *Eur. J. Lipid Sci. Technol.* 108: 134–142.

Fuster, M.D., Lampi, A.M., Hopia, A., and Kamal-Eldin, A. 1998. Effect of α- and γ-tocopherols on the autoxidation of purified sunflower triacylglycerols. *Lipids* 33(7): 715–722.

Gana, H.L., Tana, C.P., Che Man, Y.B., NorAinib, I., and Nazimah, S.A.H. 2005. Monitoring the storage stability of RBD palm olein using the electronic nose. *Food Chem.* 89(2): 271–282.

Gertz, C. 2004. Optimizing the baking and frying process using oil-improving agents. *Eur. J. Lipid Sci. Technol.* 106: 736–745.

Gertz, C. 2000. Chemical and physical parameters as a quality indicator of used frying fats. *Eur. J. Lipid Sci. Technol.* 102: 566–572.

Gil, B. and Handel, A.P. 1995. The effect of surfactants on the interfacial tension of frying fat. *J. Am. Oil Chem. Soc.* 72(8): 951–955.

Goburdhun, D., Jhaumeer-Laulloo, S.B., and Musruck, R. 2001. Evaluation of soybean oil quality during conventional frying by FTIR and some chemical indexes. *Intl. J. Food Sci. Nutrition* 52(1): 31–42.

Guadarrama. A., Rodrguez-Mendez, M.L., Sanz, C., Ros, J.L., and de Saja, J.A. 2001. Electronic nose based on conducting polymers for the quality control of the olive oil aroma — Discrimination of quality, variety of olive and geographic origin *Anal. Chim. Acta* 432(2): 283–292(10).

Hodgins, D. 1996. Electronic nose technology. *Perfumer and Flavorist* 21(3): 45–46, 48.

Hodgins, D. and Simmonds, D. 1995. Sensory technology for flavor analysis. *Cereal Foods World* 40(4): 186–191.

Holownia, K.I., Chinnan, M.S., Erickson, M.C., and Mallikarjunan, P. 2000. Quality evaluation of edible film-coated chicken strips and frying oil. *J. Food Sci.* 65(6): 1087–1090.

Hunter, J.E. and Applewhite, T.H. 1993. Correction of dietary fat availability estimates for wastage of food service deep-frying fats. *J. Am. Oil Chem. Soc.* 70(6): 613–617.

Innawong, B. 2001. Improving fried product and frying oil quality using nitrogen gas in a pressure frying system. Ph.D. Dissertation. Virginia Polytechnic Institute and State University.

Innawong, B., Mallikarjunan, P., Irudayaraj, J., and Marcy, J.E. 2003. The determination of frying oil quality using Fourier transform infrared attenuated total reflectance. *Lebensm-Wiss. u-Technol.* 37(1): 23–28.

Innawong, B., Mallikarjunan, P., and Marcy, J.E. 2003. The determination of frying oil quality using a chemosensory system. *Lebensm-Wiss. U. Technol.* 37: 35–41.

IUPAC. 1992. *Standard methods for the analysis of oils, fats, and derivatives*, 1st Supplement to the 7th ed., International Union of Pure and Applied Chemistry. Oxford, England: Pergamon Press .

Jaswir, I., Che Man, Y.B., and Kitts, D.D. 2000a. Synergistic effect of rosemary, sage and citric acid on retention of fatty acids from refined, bleached and deodorized palm olein during repeated deep-fat frying. *J. Am. Oil Chem. Soc.* 77(5): 527–533.

Jaswir, I., Che Man, Y.B., and Kitts, D.D. 2000b. Use of natural antioxidants in refined palm olein during repeated deep-fat frying. *Food Res. Intl.* 33: 501–508.

Jung, M.Y. and Rhee, K.C. 1994. Quality improvement of used frying cottonseed oil by adsorbent or chemical treatment. *Foods Biotechnol.* 3: 65–70.

Kazemi, S., Wang, N., Ngadi, M., and Prasher, S.O. 2005. Evaluation of frying oil quality using VIS/NIR hyperspectral analysis. *Agric. Eng. Intl.* 7(FP 2005).

Kita, A. and Lisinska, G. 2005. The influence of oil type and frying temperatures on the texture and oil content of French fries. *J. Sci. Food Agric.* 85: 2600–2604.

Lang, W., Sokhansanj, S., and Sosulski, F.W. 1992. Modelling the temperature dependence of kinematic viscosity for refined canola oil. *J. Am. Oil Chem. Soc.* 69(10): 1054–1055.

Lin, S., Akoh, C.C., and Reynolds, A.E. 1998. The recovery of used frying oils with various adsorbents. *J. Food Lipids* 5: 1–16.

Lin, S., Akoh, C.C., and Reynolds, A.E. 1999. Determination of optimal conditions for selected adsorbent combinations to recover used frying oils. *J. Am. Oil Chem. Soc.* 76(6): 739–744.

Lin, S., Akoh, C.C., Reynolds, A.E. 2001. Recovery of used frying oils with adsorbent combinations: Refrying and frequent oil replenishment. *Food Res. Intl.* 34(2–3): 159–166.

Matthaus, B. 2006. Utilization of high-oleic rapeseed oil for deep-fat frying of French fries compared to other commonly used edible oils. *Eur. J. Lipid Sci. Technol.* 108: 200–211.

Mielle, P. and Marquis, F. 1998. Electronic nose: Improvement of the reliability of the product database by using new dimensions. *Sem. Food Anal.* 3(1): 93–105.

Milić, B.M., Djilas, S.M., and Čanadanović-Brunet, J.M. 1998. Antioxidant activity of phenolic compounds on the metal ion breakdown of lipid peroxidation system. *Food Chem.* 61(4): 443–447.

Miller, K.S., Singh, R.P., and Farkas, B.E. 1994. Viscosity and heat transfer coefficient for canola, corn palm and soybean oil. *J. Food Proc. Preserv.* 18(6): 461–472.

Miyagi, A., Nakajima, M., Nabetini, H., and Subramanian, R. 2001. Feasibility of recycling used frying oil using membrane process. *Eur. J. Lipid Sci. Technol.* 103: 208–215.

Moreira, R.G., Castell-Perez, M.E., and Barrufet, M.A. 1999. *Deep-fat frying: Fundamentals and applications.* Gaithersburg, MD: Aspen Publishers, Inc. pp. 3–6, 33–51, 75–77, 118, 179–180.

Muhl, M., Demisch, H.U., Becker, F., and Kohl, C.D. 2000. Electronic nose for detecting the deterioration of frying fat—comparative studies for a new quick test. *Eur. J. Lipid Sci. Technol.* 102: 581–585.

Navarro, J., Obrodor, E., Carreter, J., Petschen, I., Avino, J., Perez, P., and Estrela, J.M. 1999. Changes in glutathione status and the antioxidant system in blood and cancer cells associate with tumor growth in vivo. *Free Radical Biol. Med.* 26(3/4): 410–418.

Naz, S., Sheikh, H., Siddiqi, R., and Sayeed, S.A. (2004). Oxidative stability of olive, corn, and soybean oil under different conditions. *Food Chem.* 88(2): 253–259.

Paul, S. and Mittal, G.S. 1997. Regulating the use of degraded oil/fat in deep fat/oil food frying. *Critic. Rev. Food Sci. Nutri.* 37(7): 635–662.

Peter, M. and Hakan, W. 1998. Adduct formation, mutagenesis and nucleotide excision repair of DNA damage produced by reactive oxygen species and lipid peroxidation product. *Mutation Res.,* 410(3): 271–290.

Phillips, G.O. and Williams, P.A. 2000. *Handbook of hydrocolloids.* Abington Hall: Woodhead Publishing Limited.

Phogat, S.S., Mittal, G.S., and Kakuda, Y. 2006. Comparative evaluation of regenerative capacity of different adsorbents and filters for degraded frying oil. *Food Sci. Technol. Intl.* 12(2): 145–157.

Rudnik, E., Szczucinska, A., Gwardika, H., Szule, A., and Winiarska, A. 2001. Comparative studies of oxidative of linseed oil. *Thermochimica Acta* 370(1–2): 135–140.

Shyu, S-L., Hau, L-B., and Hwang, L.S. 1998. Effect of vacuum frying on the oxidative stability of oils. *J. Am. Oil Chem. Soc.* 75(10): 1393–1398.

Tomasevic, A.V. and Siler-Marinkovic, S.S. 2003. Methanolysis of used frying oil. *Fuel Process. Technol.* 81: 1–6.

Toro-Vazquez, J.F. and Infante-Guerrero, R. 1993. Regressional models that describe oil absolute viscosity, *J. AOCS* 70: 1115–1119.

Tseng, Y.C., Moreira, R., and Sun, X. 1996. Total frying-use time effects on soybean-oil deterioration and on tortilla chip quality. *Int. J. Food Sci. Technol.* 31(3): 287–294.

Van de Voort, F.R., Ismail, A.A., Sedman, J., Dubois, J., and Nicodemo, T. 1994. The determination of peroxide value by Fourier transform infrared spectroscopy. *J. Am. Oil Chem. Soc.* 71(9): 921–926.

Van de Voort, F.R., Sedman, J. and Ismail, A.A. 1993. A rapid FTIR quality control method for determining fat and moisture in high-fat products. *Food Chem.* 48: 213–221.

Waltking, A.E. and Wessels, H. 1981. Chromatographic separation of polar and nonpolar components of frying fats. *J. Assoc. Off. Anal. Chem.* 64: 1329–1330.

White, P.J. 1991. Methods for measuring changes in deep-fat frying oils. *Food Technol.* 45(2): 75–83.

Xu, X-Q. 2003. A chromametric method for the rapid assessment of deep frying oil quality. *J. Sci. Food Agric.* 83: 1293–1296.

Yang, M.Y., Han, K.Y. and Noh, B.S. 2000. Analysis of lipid oxidation of soybean oil using the portable electronic nose. *Food Sci. Biotechnol.* 9(3): 146–150.

Yang, H. and Irudayaraj, J. 2000. Characterization of semisolid fats and edible oils by Fourier transform infrared photoacoustic spectroscopy. *J. Am. Oil Chem. Soc.* 77(3): 291–295.

Yates, R.A. and Caldwell, J.D. 1992. Adsorptive capacity of active filter aids for used cooking oil. *J. Am. Oil Chem. Soc.* 69(9): 894–897.

Yates, R.A. and Caldwell, J.D. 1993. Regeneration of oils used for deep frying: A comparison of active filter aids. *J. Am. Oil Chem. Soc.* 70(5): 507–511.

Yoon, J., Han, B-S., Kang, Y-C., Kim, K.H., Jung, M.Y., and Kwon, Y.A. 2000. Purification of used frying oil by supercritical carbon dioxide extraction. *Food Chem.* 71(2): 275–279.

5 Batter and Breading — Classification and Processing Systems

5.1 INTRODUCTION

The consumer market for deep-fried coated foods has expanded rapidly in recent years. This is due to combinations of the distinctive flavor, aroma, and crunchy texture characteristics along with the flavors and the juices that are retained in the core and in the crispy crust of products. Batter and breading are traditional methods used in food preparation to add value to products and control their texture, flavor, weight, volume, moisture loss, and fat absorption during frying. Empiricism has dominated the applications of batter coating in the food industry for decades. Consumers currently demand more sophisticated and diverse foods with less oil. As a basis for formulating appropriate batter systems, it is important to understand in detail the requirements that must be met by the batter and breading while they are raw and the characteristics they must develop on the product during and after frying.

Many studies have been published on fat uptake during frying in various battered or breaded products. Makinson, Greenfield, Wong, and Wills (1987) compared battered products with unbattered products and observed that the batter slowed penetration of fat into the food being fried. Batter and breading act as barriers against moisture loss by protecting the natural juices of the foods from the effects of freezing or reheating, thereby ensuring a final product with a tender and juicy core coated with a crispy crust.

A batter and breading system is a highly complex one in which the nature of the ingredients is very wide-ranging and interactions between ingredients determine the performance quality of the final product. The initial moisture and fat content of the substrate (chicken nuggets, meat, squid rings, pieces of vegetable, onion rings, etc.) and the shape of the product influence initial batter pickup and later oil uptake during deep-fat frying. Coatings may also prevent oxidation, limit moisture and oil transfer, give freeze/thaw stability, and extend the shelf life of the product. They must be applied to stick and stay stuck to product surfaces under wide-ranging conditions. At the same time, coatings should not stick to each other. They might require cooking, precooking and reheating, possibly combined with freezing, or none of these. In addition, they must also be cost-effective.

It is known that the surface properties of foods are very important in influencing fat uptake as well as the appearance formed during deep-fat frying. The coating provides a promising route that provides many options to develop flavor and to reduce fat uptake when applied appropriately. The mechanism of the action is usually not fully understood, although sometimes the ingredient functionality is ascribed to a specific property. As a result, research on battered products during the last decade has focused mainly on reducing the quantity of oil absorbed during frying. In this regard, certain hydrocolloids that have been incorporated into batter formulations have shown their effectiveness as oil absorption barriers (Huse et al., 1998; Rimac-Brncic, Lelas, Rade, and Simundic, 2004; Akdeniz, Sahin, and Sumnu, 2006). Modified starch, rice flour, and other ingredients also contribute to the reduction of oil content in fried coated foods (Mukprasirt, Herald, Boyle, and Boyle, 2001).

There has been rapid growth and many technological advances in recent years related to batter formulation, breading manufacture, frying oils, and production equipment engineering. As a result, many food companies now operate and fund their own research programs for batters and breadings and their applications to various food products. These developments have resulted in improved quality assurance practices and better technical services.

The primary purpose of this chapter is to document advances in the batter and breading industry. The secondary purpose is to serve as an information resource to industry, researchers, and product development personnel. This chapter is an overview of batter and breading systems and their uses in the industry. The functionality of ingredients, the role of seasonings, the preparation of a food substrate for coating, the rheology of coating systems, and the selection and maintenance of processing equipment are described in the following.

5.1.1 Definition of Batter and Breading

Batter can be defined as liquid dough, considered to be a thick but pourable mixture, basically consisting of flour or starch blended with water into which a product is dipped before it is breaded or fried. It sometimes incorporates a leavening agent, leading to expansion when it is fried. In practice, however, batters can also contain other ingredients, including salt, seasonings, gums, egg, and many other items. Therefore, batters become highly sophisticated, complex systems in which the nature of the ingredients is very wide-ranging, and their interaction determines the final performance of the product.

Breading is a flour-based breadcrumb or cracker meal that is applied to a food in a dry form, primarily to create a desired coating texture. It is a dry food coating made from flour starch, seasonings, etc. that is coarse in nature and is applied over moistened or battered food products. The coating can be fine to coarse in particle size (Suderman and Cunningham, 1983).

5.2 BATTER CLASSIFICATIONS

Batter and breading serve as coatings that may be applied separately or in combination to produce the desired effects. The main function of a batter is to provide a base so the crumbs will adhere to the product. Although different types of batter and breading may be grouped by different names, Loewe (1993) classified batter systems into two categories, namely interface/adhesion and puff/tempura.

5.2.1 INTERFACE/ADHESION BATTER

This coating is typically used with a supplemental breading or breadcrumb and is chosen for the granulation, color, flavor, and crispness desired in the finished coated food. The batter coating serves primarily as an adhesive layer between the food surface and the breading. Chemical leavening is not normally used. These batters are pumpable and can be re-circulated. Acceptability of a finished product is determined by the uniformity and thickness of the batter coating. The interface/adhesion batter achieves this by its viscosity development. Olewnik and Kulp (1993) reported that the optimum viscosity range for wheat flour-based batters, judged by the quality of coating, was within the range of 1050 to 1200 cP. Logically, a more viscous batter will yield a higher pickup of breading than would a thin, more watery suspension. Interface batters usually contain high percentages of wheat and/or cornstarch, which can be modified chemically or thermally to improve adhesive properties.

Starch-based batters are usually thin and need to be stirred frequently to keep the starch in suspension. They bind themselves quite strongly to protein-based products and can be used as an outer coating on the product in order to increase its crispness and improve holding time.

5.2.2 PUFF/TEMPURA BATTER

Both wheat and corn flours play an important role in this batter system. Chemically leavened, the batter itself can serve as the outside coating of the food and thus requires visual and structural qualities more complex than those of the interface/adhesion batter requires. Tempura batters should not be pumped or re-circulated because pumping or mixing removes the carbon dioxide. Pickup on tempura batter is between 40 and 60%, depending on desirable thickness and product surface area in relationship to its weight (Johnson and Hutchison, 1983).

5.3 BREADING CLASSIFICATIONS

The word "breading" is a general term that refers to a large group of ground flour-based coatings (Suderman and Cunningham, 1983). Breadings have wide variation in size, color, and materials to be used alone or in combination. Mostly they contain crumbs, flours, starches, spices, herbs, nuts, and seeds. Manufacturers have continued to develop a wide variety of crumbs instead of simple basic breadcrumbs to satisfy consumers' demands. Some of the breadings include the following:

1. Reclaimed breadcrumbs: This group of breadings is made from dried (sometimes toasted) and ground unsold bread. It has variable properties and it is losing its popularity.
2. Flour-based breadings: These breadings are mostly made from wheat flour. They may contain corn (maize) flour, starches, gums, coloring agents, and seasonings to enhance adhesion and flavor attributes. This type of breading formulation is common in the Shore-lunch® brand of products.

3. Cracker meal breadings: These breadings are produced from a cracker-type formulation, generally unleavened, that results in a relatively hard texture. They vary in sizes from fine (60 to 140 mesh per inch, U.S. sieve) to medium (20 to 60 mesh per inch, U.S. sieve) to coarse (4 to 6 mesh per inch, U.S. sieve).

4. American breadcrumbs (ABC): These are manufactured from dried and ground loaves of yeast-raised bread. They are typically used on raw-breaded or par-fried products for oven or fry cooking. The breadings have a crisp texture that is not as tough as cracker meal breadings.

5. Japanese, or oriental-style, breadcrumbs (JBC): These breadings are also called Panko-type breadings. They have a light flake-like appearance and are produced by induction or resistance heating rather than by conventional oven baking. They give crisp surfaces but break easily in handling and have a very attractive coating for strong visual appeal. They create a more elongated shape than ABCs and provide a porous structure that provides a crispy, but not hard, texture.

6. Extruded crumbs: These are dried ground cracker crumbs from baked cracker-like sheets that have relatively little flavor and a hard texture. They are applied to form a fine coating over a thick batter and have a high frying tolerance against deep-fat frying for extended times. They are widely used in the United States.

7. Fresh crumbs: These are made of undried bread crumbs. They give a light and crispy texture and are perishable due to moisture content. Therefore, they must be refrigerated.

8. Cereal flakes: An example of this type of breading is ground corn flakes. They have handling and functionality characteristics similar to Japanese-style breadcrumbs.

9. Potato flakes: An example of this type of breading is ground potato flakes. They also resemble Japanese-style breadcrumbs.

10. Dry flavorings and sauces: These are usually considered as color and flavor contributors.

5.4 CRITICAL COATING CHARACTERISTICS

Both batters and breadings serve many functions as food coatings such as enhancing the appearance and taste characteristics of a food product. Consumers normally purchase a battered or breaded food based on several general quality factors. Attributes that can be evaluated include color and appearance, crispness, texture, flavor, and moisture and oil content.

5.4.1 COLOR AND APPEARANCE

Appearance, including color and overall aesthetics, is largely affected by the amount and uniformity of the coating adhering to the food substrate, which depends on the properties of the food substrate, batter ingredients, and cooking methods. The remaining coating adhering to the food determines the thickness of the final layer, as well as some of its texture characteristics and, from an industrial point of view, the ultimate yield of the product (Salvador, Sanz, and Fiszman, 2002).

Suderman and Cunningham (1977) reported achieving better appearance and greater coating uniformity on fresh chicken parts than on parts that had been frozen and thawed. Hanson and Fletcher (1963) found appearance to be affected by thickening agents included in the coating mix. For example, thicker coatings formulated from waxy rice flour and corn flour were very smooth, probably too smooth to be attractive. However, thinner coatings made from the same flours allowed bubbles to appear on their surface. The bubbles became larger as coating thickness increased. Therefore, the ingredient composition does play an important role in the development of an acceptable coating appearance.

Cooked color is closely related to coating appearance. It results from a combination of ingredient composition, cooking method, coating medium, and cooking oil. Composition plays an important role in determining the extent of color development due to Maillard reactions. Therefore, protein and sugar sources are important to the final coating color. Different studies have found coating color to depend on cooking time and temperature, breading material composition, and cooking oil characteristics and composition (Landes and Blackshear, 1971).

Golden brown always has been the benchmark color for fried foods. Now restaurants are demanding more variety, and colors can be as diverse as autumn leaves. Many brown and toasted colors result from Maillard reactions. Product designers can also add caramel colors to enhance brown notes in baked applications. Retail breading mixtures frequently contain sugars and malt dextrin to hasten browning. Any reducing sugar, such as dextrose or lactose, is helpful in adding brown notes. These carbohydrates contribute appeal not only with Maillard reaction browning, but also by adding flavor through caramelization. The lactose component of whey acts as a reducing sugar and contributes to Maillard browning. The presence of milk solids in batter significantly darkens the batter, increases redness, and decreases yellowness of fried products.

Coating color can be measured with optical colorimeters that can gauge the spectral components of coating colors.

5.4.2 CRISPNESS

Crispness and texture impart various taste qualities to a coating that are important to the pleasure derived from eating fried foods. Texture is a result of its underlying structure and the mechanical properties of that structure (Vickers and Bourne, 1976). Although crispness depends upon the product composition, it can be adjusted by time and temperature of cooking. A lack of crispness may be defined as either a chewy toughness or a mushy softness. Ideally, the coating should exhibit a structure that sufficiently resists the initial bite but then disappears with a quick melt in the consumer's mouth. Indeed, crispness is a difficult attribute to maintain in a batter/breading system during the extended storage times desired by the retail processor (Loewe, 1993).

Mohamed, Hamid, and Hamid (1998) reported that crispness was positively correlated with amylose content. The addition of pregelatinized rice flour improved crispness, but resulted in increased oil absorption because of the porous nature of the fried product. The results also showed that adding calcium chloride and ovalbumin reduces oil absorption and increases batter crispness. Matsunaga, Kawasaki, and Takeda (2003) studied the influence of physicochemical properties of starch on crispness of tempura-fried batter. They found the crispness (favorable eating texture) of tempura coating depended largely on the starch origin. Experimental results indicated that amylose was one of the determinants that controlled the crispness of fried batter by restraining the disintegration of the starch granule structure.

A whey protein concentrate (WPC) with 80% protein will result in a crisper coating. Several miscellaneous ingredients also play a part in batter structure. In the food service sector, batter and breading manufacturers are bringing a completely new dimension in taste and texture to fried foods. A challenge is to keep these products crispy and crunchy while they sit under heat lamps waiting to be served.

The attributes of crispness or toughness, expressed as resistance to mastication, can be examined through compressibility tests. Here, the resistance value of the coating itself can be determined after either a fixed or variable rate of stress is applied to the sample.

5.4.3 TEXTURE

Texture is important to the overall consumer acceptability of breaded fried foods. The texture of the end product depends upon the ingredients, formulation (proper balance among ingredients), and processes such as batter mixing system, coating methods, and frying conditions. However, the ingredients used in a formula are the determining factors controlling the quality of the final product. From a sensory point of view, the important texture parameters are firmness, springiness, stickiness, etc. All these textural characteristics are perceived by the mouth and teeth, through stresses and strains imposed during deformation in the mouth. Consequently, since rheological properties also relate to the stress–strain behavior of a material, they should give information related to texture. From a physical standpoint, the coating volume is one of the most crucial parameters of the texture quality of the finished coated product.

5.4.4 FLAVOR

Flavor affects the human senses and stimulates our desire to eat or reject various foods. It depends on cooking time and temperature, composition and characteristics of the frying oil, and composition of the breading materials (Loewe, 1993). The coating seems to seal in and soak up the juices normally lost during the steam cooking process (Love and Goodwin, 1974). Although a fixed recipe of additives and seasoning may be used, flavor is still dependent on the method, time, and temperature of cooking, the composition and characteristic of the frying oil, and the type of supplemental batter system.

5.4.5 MOISTURE AND OIL CONTENT

More juices and less oil content are desirable in the final fried coated product. Processors require that batter/breading systems control moisture transfer to keep the substrate's juice content inside the coated product and prevent excess fat absorption during deep-frying.

During heating, moisture travels from the substrate, through the coating, and into the air as steam. This creates several potential problems: It can negatively affect substrate texture and eating qualities, lower finished-product yield, and change the coating texture or affect its adhesion. Hydrocolloids also help address this problem. Several research works deal with the relationship between moisture content and oil uptake, and most results claim that higher initial moisture content results in a higher fat uptake (Gamble, Rice, and Selman, 1987; Moreira, Sun, and Chen, 1997; Mellema, 2003). Lower initial moisture content presumably would lessen the internal volume of the food that could be occupied by oil during frying and would shorten the frying time. Sustaining high moisture content in the final product normally results in a low final fat content.

Makinson et al. (1987) studied the fat uptake during deep-fat frying of coated and uncoated foods. In this experiment, 20 meats and plant foods were deep-fat fried for varying times and their fat contents were determined. The results showed that total fat uptake was higher in plant foods than meat, probably because the plant foods initially had high water and low fat contents. The presence of high levels of fat in raw meat has been shown to retard evaporative moisture losses. For meat, deep-frying may produce little or no change in the total fat content. However, the capacity of deep-frying to increase the fat content of a food with low initial fat is important. Oil absorption is also affected by batter porosity and breading properties during deep-frying. Hydrocolloids and some proteins show their capacity to improve batter and breading systems by reducing oil uptake. Several miscellaneous ingredients also play factors in batter structure. It has been shown, for example, that adding calcium chloride reduces oil absorption and increases coating crispness.

The type of breading and the number of layers will affect the total fat uptake. The oil condition and temperature also influence the amount of fat absorbed during frying. Formulating batters with ingredients with film-forming properties also can result in reduced fat pickup. Fat reduction has been a major concern, but recently other concerns have centered on ensuring that healthy compounds are maximized while unhealthy compounds are eliminated. There has been much interest in adding soy flour, soy protein, and textured soy components to batters and breadings. Another possibility is the addition of omega-3 fatty acids, but this objective is not easy to achieve. The bottom line is that most consumers currently eat battered and breaded foods for the crunchy and tasty feel, and not necessarily for nutrition.

5.5 THE MARKETING OF BATTER AND BREADING PRODUCTS

Both children and adults favor battered and breaded products because a crispy and crunchy coating, whether fried or baked, adds an extra dimension of texture and flavor. There are great varieties of frozen breaded or battered products available today. These advances are demonstrated most vividly by the popularity of battered fried chicken, shrimp, scallops, oysters, and fish. Recently, appetizers have become a growing category of breaded products that includes cheese sticks and vegetables such as jalapeno peppers, mushrooms, okra, corn, zucchini, broccoli, cauliflower, and even dill pickle spears.

The market for batters and breadings is divided into two main categories: food service and retail. The food service sector, which includes restaurants and fast food chains, offers a tremendous variety of battered and breaded items including fried appetizers, entrées, and vegetables. The other side of the market is the retail sector, which primarily includes frozen products that consumers bake in their home oven. In the food service sector, batter and breading manufacturers are bringing a completely new dimension in taste and texture to fried foods. A secondary challenge is to keep these products crispy and crunchy while they sit under heat lamps waiting to be served. In the retail sector, the food formulator must create a product that tastes and crunches like something out of the fryer even though it is cooked in an oven or microwave. While baked products can come pretty close to fried products through selected modifications of the formula and process, microwave products do not turn out the same with today's technologies.

5.6   BATTER AND BREADING PROCESS EQUIPMENT

The batter and breading equipment is an important factor in the process and needs to be appropriate for the chosen batter or breading. Because the types and configuration of application systems vary widely, it is extremely important that these parameters be considered when designing the system. A product might go through one-step, simple batter application, or it might be subjected to a multi-step operation such as predust, batter, breading, batter, breading. A multi-step coating operation is usually required to achieve high levels of pickup or when the substrate is difficult to coat, as in the case of most vegetables.

The specific process and reconstitution that a product encounters affects the ingredient selection. Mechanical shear on batter can be detrimental in various ways. It can change the viscosity in the presence of ingredients that exhibit shear thinning and can alter pickup level. It can damage the starch granules and negatively affect their functionality. Shear creates heat, which can affect the leavening system, and causes breakage of the breadcrumb. There are many formula considerations when trying to make a product that runs efficiently. This section is an overview of the mechanical part of the industry.

5.6.1   BATTER AND BREADING PROCESS

The batter and breading process consists of a number of steps. The specific raw material may be in the form of a natural fillet block, deep-skinned block, or minced block. Once the raw material is selected, the size of the portion is determined. Products cut from blocks are exact portions and may be cut in a variety of geometrical shapes. Additionally, products that are formed, i.e., nuggets and shapes, allow for exact portion sizing. Products cut from natural fillets have a wider finished weight range.

A wide variety of coating and packaging options are available. Unbreaded items are ideal for grilling, baking, or batter-ready applications. Raw breaded items are designed to be deep-fried only for quick preparation. Oven-ready or "par-fried" portions may be prepared by frying or baking in the oven. Battered items are partially fried to set the coating and may be prepared by frying or cooking in the oven. Flavor glazing is done to add additional flavor. Glazed products are typically used for grilling, baking, or broiling applications. A variety of coatings may be used for different applications and desired flavor profile. Packaging is engineered according to the product, its application, and customer needs. Figure 5.1 shows four basic coating systems (Johnson and Hutchison, 1983).

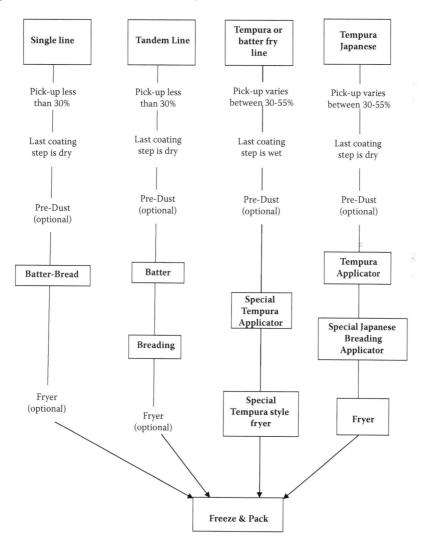

FIGURE 5.1   Four types of coating systems.

5.6.2 Characteristics of Process and Equipment

In general, a system is custom designed to account for production differences and finished product preferences. The mechanical system used for coating depends on the kind of substrate that will be coated and the method that will be used for the coating. If a product has traditionally been dipped once into a batter and then breaded once, the mechanical steps follow the same procedure and involve a single batter-breading machine combination. If the product, such as a fish fillet, is normally predusted or given a light dusting first and then batter-breaded, then the equipment components are committed to the same sequence. Each coating step, whether it involves a wet or dry material, requires a separate machine or component. The individual units are positioned in a system to accomplish a specific coating step by a continuous process, duplicating the traditional manual operation.

There are two basic types of batter applicators (Figure 5.2), both for applying wet coatings, and two types of breading machines for applying dry materials. One such breading applicator is shown in Figure 5.3. The variety of combinations of coating equipment designs is numerous, as is evident in Table 5.1 through Table 5.3 (Sebranek, 2004).

Submerger

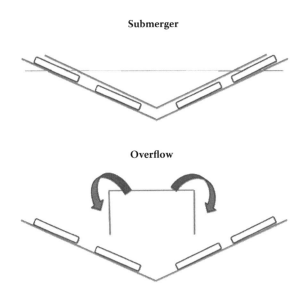

Overflow

FIGURE 5.2   Batter application methods.

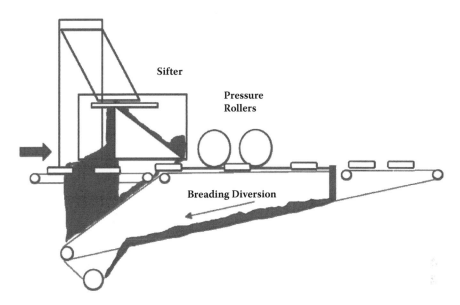

FIGURE 5.3 A schematic representation of a breading applicator (the dark areas represent breading).

TABLE 5.1
Basic Machining Characteristics of Batters

Batter Type	Mixing	Temperature	Viscosity Control
Non-leavened conventional	Can be pumped. Batters are mixed continuously to keep batter solids in solution. Viscometer cup reading to 25 sec.	Preferably under 10°C (40–50° F)	It is possible to obtain a continuous viscosity control. The Stein ABC unit automatically mixes a dry batter mix with water to a pre-set thickness (viscosity) and maintains that thickness continuously. It also controls the batter temperature.
Leavened (Tempura)	Cannot be pumped. Mixing must be done quickly and then stopped. After that, batter should be transferred quickly to the applicator.	Cold batter: 4–7°C (40–45°F) Warm batter: 18–24°C (65–75°F)	There is no continuous viscosity control. Identical batches only.

TABLE 5.2
Basic Machining Characteristics of Breading

Breading Type	General Characteristics	Machine Handling	Machine to Be Used
Free-flowing breading	Fine, uniform granulation, not fragile.	Easy to handle, flows without problems in machinery.	Can be run on every machine.
Flour-type breading	Usually raw flour; can be a dry batter mix.	Difficult to handle, material packs and bridges easily. This type of breading must be driven (pushed through the machine).	Special type of machine must be used. Usually can run free-flowing coatings as well.
Coarse breading (Japanese, Panko, Oriental)	Non-uniform granulation, fine and coarse particles.	Extremely difficult to handle, presents two challenges to the processor. 1. Breading must be prevented from grinding up 2. Product must be uniformly coated on top and bottom with the same ratio of fine and coarse crumbs	Only a special type of machine can be used. Usually can run free-flowing coatings as well.

TABLE 5.3
Other Varieties of Breading

Breading Type	Machine to Be Used
Green bread crumbs	Bread that has been ground but not dried. This type of manufacture represents the first step in processing Japanese breadcrumbs prior to drying them. They usually are mixed with free-flowing, fine-type breading material in order to run in a breading machine.
Potato flakes	These are flaky and somewhat similar to Japanese breadcrumbs.
Cereals	Ground corn flakes are an example. They have handling characteristics similar to the coarse or Japanese-style bread crumbs particularly with regard to machinability.
Cracker crumbs	Cracker crumbs have a considerable variation in machining characteristics. Generally, they will run on a machine that handles free-flowing fine reading.

5.7 APPLICATION OF BATTERS AND BREADING

A wide variety of foods, vegetable, poultry, seafood, and red meat can be battered and fried. Each substrate has special characteristics, such as cell structure, size, and cooking parameters. Coatings are specialized for each type of food and paired with flavors that blend well with each substrate. Coatings should be modified based on the desired finished product. For example, chicken does not contain as much water as fish, so chicken batters need more protein and gluten to bind the batter to the meat or skin. For a fish substrate, more starch is needed to bind the extra water.

5.7.1 VEGETABLES AND CHEESES

Onions are the most commonly encountered coated vegetable. Other battered or breaded vegetable products are bell peppers, cauliflower, eggplant, mushrooms, okra, and zucchini. Burge (1983, 1990) demonstrated the primary type of coating system used for sliced onion rings to follow these sequences: predusting, batter, breading, batter, breading. Predusting is optional, and provides improved first batter adhesion on the rings. Without predusting, batters that are usually mixed slightly thicker than whole milk will tend to run off the rings. The two passes of batter and breading provide the amount of coating desired while avoiding too much batter or breading in any one layer of coating.

Vegetables create unique coating problems because of their own specific shapes and textures and their extremely high moisture content. Their smooth waxy surfaces create problems in getting the batter to adhere properly (Suderman, Wiker, and Cunningham, 1981). Makinson et al. (1987) reported that vegetables have higher oil absorption than meat during deep-fat frying. However, vegetables still make great candidates for breading. Recent additions to battered and breaded lines include rice balls, risotto, and potato balls. Fried, breaded cheese sticks are popular appetizer items that require special restricted-melt cheeses. Mozzarella is the most popular cheese for frying, but cheddar and American cheese fillings are often sandwiched into fried vegetables and entrées. Products such as cheese sticks utilize a triple-pass system (layers will likely include batter, another batter, and then a breading) to eliminate the risk of blowout. A blowout can result when water causes a negative exchange of moisture and fat, resulting in coating separation.

5.7.2 SEAFOOD

Breaded seafood represents another significant portion of battered and breaded foods. Breaded seafood may be battered, crumbed, or both. When frozen in ready-to-cook form, they offer a convenience food valued widely by the consumer. Fish fingers, fish portions, and fish cakes are the staple breaded seafood lines, while breaded oysters, shrimp, and scallops cater to a luxury market and are widely used by the restaurant trade. Seafood processors can lessen the batter/breading adhesion problems by melting the ice layer with a surface application of salt. A flour-based predust can then absorb the surface moisture. However, an excessive level of predust can exacerbate adhesion problems. Most systems use consecutive layers of predust, batter, and breading to achieve the desired textural or functional results.

Two popular batters for seafood, designed to be used without breading, are tempura and fish-and-chip batters. These are flour-and-starch-based batters that contain leavening agents to create a more open texture. The combination of sodium bicarbonate and a leavening acid forms carbon dioxide bubbles trapped in the matrix when it is heated. Batters are used on fish, either in conjunction with breading or alone, to create a more dough-like texture. Adhesion batters are designed to provide an adhesive layer between the substrate and the outer breading layer. They can be formulated with wheat flour, corn flour, or starch to provide viscosity during the application process and to hold moisture.

For shrimp and similar products, manufacturers generally use JBC, also known as panko. These are prepared from bread dough baked in a special oven where an electric current passes through the dough. This produces a porous crumb with a sliver-like texture and no visible crust color. JBC produces a very light and crispy breading that works well on many substrates, especially shrimp. Though light in texture, JBCs have been improved to be resilient and to stand up well to machining.

The technology behind basic fish sticks remains relatively unchanged from its earlier years. Fresh, boneless fish is frozen in block form, making a raw material known as fish block. When ready to process, these large bricks of between 16 and 40 lb are run through a series of band saws and chopping mechanisms that shape them into generally rectangular portions. The weight of these portions typically ranges from 2 to 4 oz.

Because fish and seafood fall under the jurisdiction of the FDA rather than the USDA, these products have no restrictions on the level of coating applied. This can be an extreme advantage for adding value, or more specifically, profitability, for the producers. Carbohydrate-based batters and breadings are relatively inexpensive compared to protein-based seafood. Higher application rates, particularly on expensive substrates like larger shrimp, can bring down the finished-product cost per pound significantly. Typical batter coating levels go as high as 40 to 60%, according to the marketing strategy. These high levels can lower the cost, but may raise the bar in terms of technology (Gerdes, 2001).

5.7.3 Poultry

The market for battered and breaded chicken pieces exploded in the 1980s. Presently, poultry takes the lion's share of the market in the United States (in Europe, coated fish has the biggest market share). The fast-food industry continues to capitalize on chicken nuggets and strips with the trend toward more texture in the coatings. Flavors include honey-battered, buttermilk, savory, and spice marinade (Parinyasiri, Chen, and Reed, 1991).

The second most popular category in the United States is appetizers such as chicken "drummies." Formed pieces of chicken without bone are hot items in the appetizer line. The popular jalapeño poppers are also hot in more ways than one. Restaurants are looking for unique signature items and some companies now offer a line of cold-form extrusion equipment, allowing creation of distinctive appetizers. The technology co-extrudes a sauce or filling on cold-form extrusion equipment.

The structure of poultry skin is important in understanding the adhesion of batter and breadings. It consists of two layers: the epidermis and the dermis. The skin acts as a two-way barrier against the entrance of pathogens and the escape of vital fluids and gases (Lucas and Stetterheim, 1972; Korver and Klasing, 2001). The effect of freezing poultry parts before applying batters and breadings is important because freezing is a common consumer and industry practice (Cunningham, 1983).

5.7.4 RED MEAT

For meat, a major objective is to retain moisture during frying to increase cook yields and improve processing. Maintaining a high moisture gradient between the inside and the outside helps achieve this goal. Using polyphosphate can increase moisture retention, especially in fish blocks and similar products. It is believed that polyphosphates cause meat proteins to swell and increase their water-binding capacity. High-pH phosphates also can affect the flavor of mild-flavored meats, such as seafood and poultry, so careful selection of polyphosphates will reduce the risk of off-flavors. Breaded beef products include beef-fried steak, beef fritters, and beef fingers. These include single whole muscle, shaped whole muscle, and patty forms. Similar pork and veal products also are available. Popular varieties include regular, extra crispy, and Southwest style. The USDA limits the level of batter and breadings on red meat and poultry to no more than 30% of the finished product weight. Two exceptions are corn dogs and fritters, which can carry up to 65% coating.

5.8 REFERENCES

Akdeniz, N., Sahin, S., and Sumnu, G. 2006. Functionality of batters containing different gums for deep-fat frying of carrot slices. *J. Food Eng.* 75(4): 522–526.

Burge, R.M. 1983. Application of batters and breadings to onion rings and other vegetables. In: *Batter and breading technology.* Suderman, D.R. and Cunningham, F.E. (Eds.). Westport, CT: AVI Publishing Company.

Burge, R.M. 1990. Functionality of corn in food coatings. In: *Batters and breading in food processing.* Kulp, K. and Loewe, R. (Eds.). St. Paul, MN: Association of Cereal Chemists, Inc.

Cummings, G. 1983. The facts of frying. *Restaurant Bus.* 82(6): 246, 248, 250.

Gamble, M.H., Rice, P., and Selman, J.D. 1987. Relationship between oil uptake and moisture loss during frying of potato slices from c.v. Record U.K. tubers. *Intl. J. Food Sci. Technol.* 22(3): 233–241.

Gerdes, S. 2001. Batters and breadings livens tastes. *Food Product Design.* http://www.food-productdesign.com/articles/462/462_1201de.html. Accessed June 18, 2008.

Hanson, H.L. and Fletcher, L.R. 1963. Adhesion of coatings on frozen fried chicken. *Food Technol.* 17: 115.

Huse, H.L., Mallikarjunan, P., Chinnan, M.S. et al. 1998. Edible coatings for reducing oil uptake in production of akara (deep-fat frying of cowpea paste). *J. Food Process Preserv.* 22(2): 155–165.

Johnson, R.T. and Hutchison, J. 1983. Batter and breading processing equipment. In *Batter and breading*, Vol. 1. Suderman, D.R. and Cunningham, F.E. (Eds.). Westport, CT: Avi Publishing Company, pp. 120–154.

Korver, D. and Klasing, K. 2001. Influence of nutrition on immune status of the bird. Proceedings of the 24th Technical Turkey Conference, Leyburn, England, April 26–27, 2001, p. 43.

Landes, D.R. and Blackshear, C.D. 1971. The effects of different cooking oils on flavor and color of fried chicken breading material. *Poultry Sci.* 50: 894–897.

Loewe, R. 1993. Role of ingredients in batter systems. *Cereal Foods World* 38(9):673–677.

Love, B.E. and Goodwin, T.L. 1974. Effects of cooking methods and browning temperatures on yields of poultry parts. *Poultry Sci.* 4: 1391–1398.

Lucas, A.M. and Stetterheim, P.R., 1972. Avian Anatomy Integument Part I and II, Agricultural Research Service, USDA, Washington, D.C.

Makinson, J.H., Greenfield, H., Wong, M.L., and Wills, R.B.H. 1987. Fat uptake during deep-fat frying of coated and uncoated foods. *J. Food Comp. Anal.* 1: 93–101.

Matsunaga, K., Kawasaki, S., Takeda, Y. 2003. Influence of physicochemical properties of starch on crispness of tempura fried batter. *Cereal Chem.* 80(3): 339–345.

Mellema, M. 2003. Mechanism and reduction of fat uptake in deep-fat fried foods. *Trends Food Sci. Technol.* 14(9): 364–373.

Mohamed, S., Hamid, N.A., and Hamid, M.A. 1998. Food components affecting the oil absorption and crispness of fried batter. *J. Sci. Food Agric.* 78(1): 39–45.

Moreira, R.G., Sun, X., and Chen, Y. 1997. Factors affecting oil uptake in tortilla chips in deep-fat frying. *J. Food Eng.* 31: 485–498.

Mukprasirt, A., Herald, T.J., Boyle, D.L., and Boyle, E.A.E. 2001. Physicochemical and microbiological properties of selected rice flour-based batters for fried chicken drumsticks. *Poultry Sci.* 80(7): 988–996.

Olewnik, M. and Kulp, K. 1993. Factors influencing wheat flour performance in batter systems. *Cereal Foods World* 38(9): 679–684.

Parinyasiri, T., Chen, T.C., and Reed, R.J. 1991. Yields and breading dispersion of chicken nuggets during deep-fat frying as affected by protein content of breading flour. *J. Food Process. Preserv.* 15: 369–376.

Rimac-Brncic, S., Lelas, V., Rade, D., and Simundic, B. 2004. Decreasing of oil absorption in potato strips during deep fat frying. *J. Food Eng.* 64(2): 237–241.

Salvador, A., Sanz, T., and Fiszman, S. 2002. Effect of corn flour, salt, and leavening on the texture of fried, battered squid rings. *J. Food Sci.* 67(2): 730–733.

Sebranek, J.G. 2004. Semidry fermented sausages. In *Handbook of food and beverage fermentation.* Edited by Y.H. Hui, L. Meunier-Goddik, A.S. Hansen, J. Josephson, W-K. Nip, P.S. Stanfield and F. Toldra. p. 385–396, New York: Marcel Dekker.

Suderman, D.R., Wiker, J., and Cunningham, F.E. 1981. Factors affecting adhesion of coating to poultry skin—effect of age, method of chilling, and scald temperature on poultry skin ultrastructure. *J. Food Sci.* 45(3): 444–448.

Suderman, D.R. and Cunningham, F.E. 1983. *Batter and breading technology.* Westport, CT: AVI Publishing.

Suderman, D.R. and Cunningham, F.E. 1977. Adhesion and uniformity of coating of a commercial breading mix in relation to skin ultrastructure. *Poultry Sci.* 56: 1760.

Vickers, Z.M. and Bourne, M.C. 1976. Crispness in foods — a review. *J. Food Sci.* 41: 1153–1157.

6 Properties of Batters and Breadings

The main purpose for applying bread and batter coatings on fried foods is to produce high-quality products in terms of visual appeal, crispness, low fat content, flavor, and favorable consumer satisfaction (Loewe, 1990, 1993; Fiszman and Salvador, 2003). The various ingredients used in batter and breading formulation impart different functionalities and properties to the batters (Xue and Ngadi, 2006, 2007). Selection of batters in the industry has been largely empirical. Processors have relied on trial-and-error approachs or on experience to match batter functionality with a specific application. Due to increasing consumer demands for higher quality coated products and in order to optimize the development of novel batters for industrial applications, it has become vital to understand the functional properties of batters and the influence of their ingredients. Knowledge of thermophysical properties of batters such as viscosity, particle-size distribution of breading and flour, adhesion, and porosity can be helpful for the development of new batter formulations.

6.1 RHEOLOGICAL PROPERTIES OF BATTERS

Batter rheology is the most critical batter processing parameter because it determines its pasting characteristics as well as its performance during deep-fat frying by influencing adhesion and pickup. Adhesion is defined as the chemical and physical binding of a coating both with itself and with the food product (Corey, Gerdes, and Grodner, 1987). Batter and breading loss causes major problems within the fried food industry. Also known as the "blow-off" phenomenon, batter and breading not only results in loss of material, it hastens oil deterioration during frying (Parinyasiri and Chen, 1991). This is due to the small breading pieces that are released into the frying oil which become overheated at the high frying temperature, further contributing to oil quality degradation and to burnt flavor.

Batter pickup refers to the amount of batter adhering to a food substrate during battering and is expressed as a percentage of the total product weight (Mukprasirt, Herald, Boyle, and Boyle, 2001). Batter pickup is regulated by the USDA and varies for different product types. The amount of batter pickup also has an effect on the amount of breading pickup on the outer surface of the coated food product. The appearance and texture of the final product may be compromised if there is inadequate breading pickup. Corey et al. (1987) showed that frozen storage time had the greatest effect on breading loss, i.e., the longer the storage time, the greater the breading loss.

6.1.1 BATTER VISCOSITY

Viscosity is the measurement of the resistance to flow. Batter viscosity is a major factor in determining whether a product can be coated successfully. It influences the quantity and quality of batter pickup, appearance and texture, and the handling property of the coated product (Mukprasirt, Herald, and Flores, 2000). It is affected by batter temperature, ingredient composition, and the solid-to-water ratio (Flores et al., 2000). The most common method for adjusting viscosity of batter is by changing its solids-to-water ratio (Suderman, 1990). A batter temperature between 4.5 and 15.5°C (40 and 60°F) is recommended for optimum batter application while not favoring rapid growth of microorganisms in the batter. Temperature increase normally results in increase or pasting of batter systems. Structural changes occur during heating of batter corresponding to the transition from a fluid to a gelled state, resulting in dramatic changes in viscosity. Starch granules disrupt their crystalline structure during heating, and begin to absorb water and swell with a consequent increase in viscosity. The peak viscosity of gelatinized starch is reduced by the action of α-amylase, which disrupts the starch granules during the frying operation. The swollen starch granules eventually provide a film barrier that inhibits oil penetration into the food substrate, and prevents water loss from the substrate. Gelatinization and the film formed play a significant role in providing crispness and texture to the finished fried product.

Traditionally, the Brabender Visco Amylograph is used to monitor viscosity changes of batter systems. Naruenartwongsakul, Chinnan, Bhumiratana, and Yoovidhya (2004) used the Brabender Visco Amylograph to evaluate pasting properties of wheat-flour-based batters. The amylograms from the equipment were described in terms of pasting temperature (PT), peak viscosity (PV), viscosity (V) at 95°C (V 95°C), and viscosity after a holding period (V 95°C H). The viscosity values were reported in terms of Brabender units (BU). A typical amylogram of wheat-based batters is shown in Figure 6.1. PT is the first deflection of temperature as the curve begins to rise. This parameter is affected mostly by the initial concentration of starch in the batter system. As the concentration of starch increases, lower temperatures will be required to yield viscosity sufficiently high enough to be recorded by the amylograph (Rasper, 1980). Mukprasirt et al. (2002) reported increase in peak viscosity when the ratio of rice to corn flour increased in a mixed flour batter system. This was attributed to differences in their particle sizes (rice flour granules are larger than corn flour granules) and the state of their starch.

Wheat flour is one of the major ingredients used in formulating various batters for deep-fat fried products. Olewnik and Kulp (1990) studied factors affecting performance characteristics of wheat flour in batters. Those factors were type of flour (soft or hard), protein content, damaged starch, supplementation of flours with amylolytic (cereal and fungal) and proteolytic enzymes, and the addition of oxidants and reducing agents. The protein content and the levels of damaged starch in soft wheat flour were lower than those of hard wheat flour. Flours with higher protein levels produced batters with increased viscosity and required a higher amount of water to reach the optimum viscosity (1050 to 1200 cp) compared to flours with lower protein. Flours with higher levels of damaged starch increased the viscosity of batter in both types of flour. Cereal amylase had no significant effect on both batter viscosity and pickup,

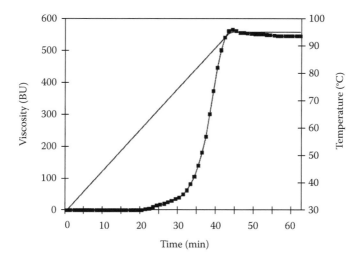

FIGURE 6.1 Typical amylogram of wheat-based batter. (Source: Naruenartwongsakul, S., Chinnan, M.S., Bhumiratana, S., and Yoovidhya, T. 2004. Pasting characteristics of wheat flour-based batters containing cellulose ethers. *Lebensm.-Wiss. u.-Technol.* 37: 489–495.)

while fungal amylase slightly decreased the viscosity of batter containing soft wheat flour but had no effect on the viscosity of batters from hard wheat flours. On the other hand, fungal protease increased the viscosity of batter containing soft wheat as compared to that containing hard wheat flour. Oxidizing agents (potassium bromate and potassium iodate) had no significant effect on the viscosity and pickup of batters, whereas reducing agents (L-cysteine and sodium bisulfite) reduced the viscosity and pickup in both hard and soft wheat flour batters.

Prakash, Haridasa Rao, Susheelamma, and Prabhakar (1998) reported that steaming wheat flour may improve the texture and the overall quality of the deep-fat fried coated products. The authors studied the flow behavior of batters (30, 33, and 36% solid concentration) made from native and wheat flour that had been steamed for 5, 15, and 30 mins. The batters behaved as pseudoplastic materials with yield stress. Steamed wheat flour was used to improve the rheological and textural characteristics of batters by induced denaturation of gluten in the batter during heat processing (Prakash and Rajalakshmi, 1999). The steamed wheat flour was used to make batters with 30, 33, and 36% solid suspension in water. The apparent viscosity of the batters increased with increase in solid concentration and steaming periods.

Processing history could affect viscosity of batters. The apparent viscosity of steamed flour was different from that of raw flour batter (Prakash, Ravi, and Susheelamma, 2001). The authors reported that apparent viscosity of a 15-min steamed flour batter was lower than that of raw flour. This result was in contrast to their findings in a previous experiment (Prakash and Rajalakshmi, 1999). Ingredients such as salt and oil also affected the rheological constants (yield stress, consistency index, and flow behavior index) of batters made from both types of flour.

The rheological constants did not change in the same manner; they depended on the combination effects of the additives, types of flour, and solid concentration.

Oliver and Sahi (1995) studied the effects of wheat cultivars (Haven, a hard milling type, and Riband, Galahad, Galahad 7, and Beaven, a soft milling type) on the rheological properties of wafer batter. Protein contents in all wheat varieties were in the range of 9.7 to 11.8%. Among the varieties studied, Haven was a hard milling cultivar and had the highest level of damaged starch. Batter made from the Haven variety had the highest rheological parameters (G', G'', and η) due to the highest amounts of damaged starch, whereas, batters made from Riband had the lowest values of these parameters. This was attributed to soft milling characteristics of the flours. The viscosity of all batters decreased with increasing shear rate.

The age of the flour or of the wheat used in making the flour may influence its viscosity. Shelke, Hoseney, Raubion, and Curran (1992) investigated the effects of flour age (0 to 14 days) and wheat age (0 to 12 weeks) on the viscosity of cake batters made from soft red winter wheat. It was found that the viscosity of cake batters increased as a function of the wheat flour age. Batter made from chlorinated flour had higher viscosity than that of untreated flour batter. In addition, the apparent viscosity of cake batter made from chlorinated flour increased rapidly during the first three days after milling, then reached a plateau after 14 days.

Besides wheat flour, rice and corn flours can also be used in batter for fried foods. Shih and Daigle (1999) stated that the addition of rice flour into wheat flour batter provided lower batter viscosity and less fluffiness in the fried batter as well as greater brittleness and hardness. Compared to wheat flour, rice flour resisted oil absorption better but was less effective as a thickening agent. Mukprasirt et al. (2000) studied the rheological characterization of rice flour-based batters (RFBB) with different ratios of rice to corn flours (50:50, 60:40, and 70:30). The rheological properties and flow behavior of RFBB were dependent upon composition, temperature, and shear rate. All RFBB formulations showed pseudoplastic behavior with a yield stress that fitted the Herschel-Bulkley model. In their study, rice flour had higher amounts of damaged starch and protein content than corn flour, which resulted in higher yield stress and consistency index in the batter containing 70% rice flour. However, the rice-to-corn flour ratio did not significantly affect the flow behavior index of RFBB.

Xue and Ngadi (2006) also studied the rheological properties of batter formulated with different flour combinations. Figure 6.2, Figure 6.3, and Figure 6.4 show viscosities of batters formulated using blends of wheat and corn, wheat and rice, and corn and rice, respectively. For the wheat and corn flour blends, batter viscosity decreased with an increasing proportion of corn flour in the batter, suggesting that corn flour diluted the strengthening influence of wheat flour gluten. Rice flour also exerted a diluting effect on wheat flour gluten, increasing the available free water in the batter system. This free water could lubricate particles, enhance flow, and result in a lower viscosity value (Mukprasirt et al., 2000). Adding rice or corn flour to substitute wheat flour in batter formulation decreased viscosity because of the dilution effect of the two flours by increasing the available free water in the batter system. Corn flour tends to present stronger influence than rice flour on viscosity in a corn-rice batter system. Addition of corn caused greater viscosity reduction than rice flour.

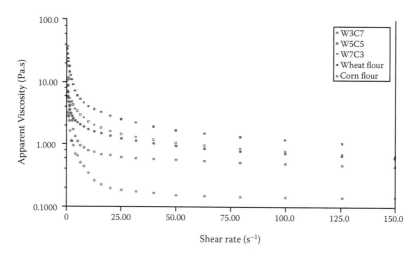

FIGURE 6.2 Viscosity of a batter system formulated with blends of wheat and corn flours. *Source*: Xue, J. and Ngadi, M. 2006. Rheological properties of batter systems formulated using different flour combinations. *J. Food Eng.* 77(2): 334–341.

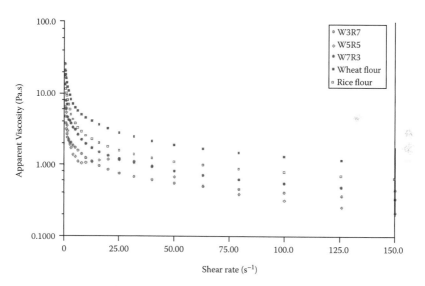

FIGURE 6.3 Viscosity of a batter system formulated with blends of wheat and rice flours. *Source*: Xue, J. and Ngadi, M. 2006. Rheological properties of batter systems formulated using different flour combinations. *J. Food Eng.* 77(2): 334–341.

The addition of some optional ingredients such as egg, starches, salts, and gums may alter the rheological characterization of batters. The addition of salt significantly lowered batter viscosity. The addition of dextrin (0 to 7.5%) into wheat flour-based tempura batters decreased the consistency index at higher concentration, whereas the flow behavior index, in general, was quite similar to those of batters without dextrin. On the other hand, the presence of dried whole egg (0 to 7.5%) increased the

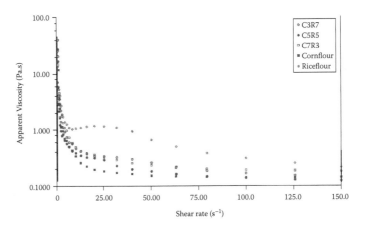

FIGURE 6.4 Viscosity of a batter system formulated with blends of corn and rice flours. *Source*: Xue, J. and Ngadi, M. 2006. Rheological properties of batter systems formulated using different flour combinations. *J. Food Eng.* 77(2): 334–341.

consistency index and decreased the flow behavior index of batter (Baixauli, Sanz, Salvador, and Fiszman, 2003). Salvador, Sanz, and Fiszman (2002) found that the addition of salt (5.5%) to wheat flour-based tempura batter lowered the viscosity and pickup of batter. Shih and Daigle (1999) investigated the effects of additives (rice starch and phosphorylated starch ester with 0.44% and 1.60% phosphorus, respectively) on batter viscosity. Batter with an equal ratio of rice and wheat flours was used as the control. The addition of unmodified rice starch (5 to 15%) did not affect the batter viscosity, whereas the phosphorylated starch increased the batter viscosity significantly. Increased phosphorus content in phosphorylated starch yielded higher batter viscosity. Moreover, Mukprasirt et al. (2000) studied the effects of oxidized cornstarch (0, 5, and 15% on a dry weight basis) and methylcellulose (0 and 0.3% on a dry weight basis) on the rheological characterization of RFBB with different ratios of rice to corn flours (50:50, 60:40, and 70:30) at 5, 15, and 25°C. RFBBs showed pseudoplastic behavior with yield stress. The addition of oxidized starch in batter decreased yield stress because starch absorbed less water than flour did, which probably caused more free water available in batter resulting in a lower yield stress. The yield stress of batters also decreased with the temperature because starch granules were not soluble in cold water. The addition of methylcellulose decreased the yield stress due to its interfacial activity, which made batters more smooth and slimy than batters without methylcellulose. The consistency index of batters did not show any remarkable changes with the addition of oxidized starch. On the other hand, addition of methylcellulose increased the consistency index, especially at the lower temperature. The addition of oxidized starch did not have a significant effect on the flow behavior index of RFBB, whereas methylcellulose decreased the flow behavior index of batters. The flow behavior index of batter also increased with increasing temperature. The type and concentrations of cellulose influenced the PV values of wheat flour-based batters (Naruenartwongsakul et al., 2004). The presence of cellulose ethers increased the PV values of the batters as shown in Figure 6.5. This may

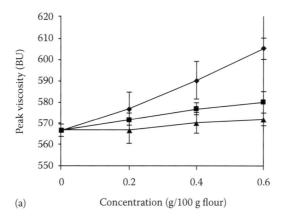

(a)

FIGURE 6.5 Effect of different concentrations of cellulose ether on peak viscosities of wheat-based flours. *Source*: Naruenartwongsakul, S., Chinnan, M.S., Bhumiratana, S., and Yoovidhya, T. 2004. Pasting characteristics of wheat flour-based batters containing cellulose ethers. *Lebensm.-Wiss. u- Technol.* 37: 489–495.

be attributed to at least two phenomena, namely the interaction of the hydrocolloid with exudate from the starch granule (solubilized amylose and low molecular weight amylopectin) and the thickening capacity of the gum.

Meyers and Conklin (1990) developed methods for reducing oil absorption in coated fried foods using HPMC with 27 to 30% methoxyl substitution and 4 to 12% hydroxypropyl substitution. The batter formulations were 45% commercial batter, 54.5% water, and 0.5% HPMC. Two techniques of adding HPMC, dry blend with solid ingredients and fully hydrated solution, were studied. HPMC increased the viscosity of batter. Further, the use of a fully hydrated HPMC solution caused the viscosity to increase significantly compared to when a dry blend was added. Hsia, Smith, and Steffe (1992) studied the effects of three hydrocolloids, namely guar, xanthan, and carboxyl methylcellulose (CMC) at concentrations of 0.25, 0.5, and 1.0% while keeping a constant batter solid content of 30%, on rheological properties and adhesion characteristics of flour-based batters for chicken nuggets. Apparent viscosities of the batters were measured using a mixer viscometer. The apparent viscosity of all batters except the control dropped when mixing time was increased, indicating thixotropic behavior. Batter mixed with xanthan gum had a higher apparent viscosity than batter mixed with other hydrocolloids, whereas batter with CMC had the lowest viscosity. The viscosity of batter increased with an increase in gum concentration, thus affecting batter pickup and yield of nuggets. The higher the batter viscosity was, the greater the pickup and yield of the batter. The work of Hsia et al. (1992) neglected the yield stress of batter. The flow behavior indices of all batters were lower than one, which indicated that the batters were pseudoplastic in nature. With the exception of CMC, the flow behavior index decreased with an increase in hydrocolloid concentration, whereas the consistency coefficient increased when gum concentration was increased.

Ang (1993) investigated the effect of fiber length (35, 65, 110, and 290 μm) of powdered cellulose (PC) on oil absorption of fried batter coatings. The results showed that the fat content of batters containing PC in excess of 100 μm decreased significantly. The viscosity of batters containing PC with length longer than 65 µm was higher than the control and batter containing 35 µm PC. The longer the fiber length, the greater the batter viscosity was. In order to identify the factor that reduced oil absorption in fried batter coating, the three gums (namely, guar gum, xanthan gum, and CMC) mentioned earlier were used. All gums increased the viscosity of batter, but did not reduce the oil absorption in fried batter. Therefore, the reduction of fat in batter coating containing PC was not due to the increase in batter viscosity.

The solid content of a batter has a direct bearing upon its functionality. Cunningham and Tiede (1981) studied the effect of batter viscosity on breading pickup and cooking loss by varying the water-to-solids ratio (2:1, 1.5:1, 1:1, 1:1.5, and 1:2). It was found that there was only a slight change in viscosity until the water-to-solid ratio reached 1:1, then the viscosity increased sharply. In addition, the breading, in the case of pickup with chicken drumsticks, increased with the increase in batter viscosity. The cooking loss also decreased as the viscosity increased. Excessively high solid content resulted in a poor weir curtain, tailing, and an objectionably tough coating. Vegetable gums added to the batter allow the reduction of the solid content, but care must be taken that the breading system is capable of coping with the additional moisture in the system (Sasiela, 1990).

Depending on the range and composition, temperature may exhibit inverse effects on batter viscosity. As a batter warms, its viscosity may decrease. Baixauli et al. (2003) studied flow behavior of tempura batter (85.4% wheat flour, 6% corn flour, 5.5% salt, and 3.1% leavening) at different temperatures of 5 to 40°C. The authors reported that the consistency index of the batter decreased with an increase in temperature. Mukprasirt et al. (2000) reported similar results in the temperature range of 5 to 25°C for rice flour-based batters.

6.1.2 VISCOELASTIC BEHAVIOR OF BATTER DURING GELATINIZATION

Viscoelastic behaviors of many foods have been studied by means of dynamic shear, creep compliance, and stress relaxation techniques (Rao, 1999). In a dynamic rheological experiment, a sinusoidal oscillating stress or strain with a certain frequency (ω) is applied to the material and the phase difference between the oscillating stress and strain, as well as the amplitude ratio, is measured. In a dynamic shear experiment, a food sample is subjected to a small sinusoidal oscillating strain or deformation $\gamma(t)$ at time t according to Equation (6.1).

$$\gamma(t) = \gamma_o \sin(\omega t) \qquad (6.1)$$

where γ_o is the strain amplitude and ω is the angular frequency. The applied strain generates two stress components in a viscoelastic material: an elastic component in line with the strain and a 90° out of phase viscous component. Differentiation of Equation (6.1) yields the strain rate $\dot{\gamma}(t)$ for evaluating the viscous component to be $\pi/2$ radians out of phase with the strain.

$$\dot{\gamma}(t) = \gamma_o \omega \cos(\omega t) \qquad (6.2)$$

For deformation within the linear viscoelastic range, the generated stress (σ_o) in terms of elastic or storage modulus G′ and a viscous or loss modulus G″ is expressed in Equation (6.3).

$$\sigma_o = G'\gamma_o \sin(\omega t) + G''\gamma_o \cos(\omega t) \tag{6.3}$$

For a viscoelastic material, the resultant stress is also sinusoidal but shows a phase lag of δ radians when compared with the strain. The phase angle δ covers the range of 0 to π/2 as the viscous component increases. The sinusoidal variation of the resultant stress is shown in Equation (6.4).

$$\sigma(t) = \sigma_o \sin(\omega t + \delta) \tag{6.4}$$

The following expressions that define viscoelastic behavior can be derived from Equation (6.3) and Equation (6.4).

$$G' = \left[\frac{\sigma_o}{\gamma_o}\right]\cos\delta \tag{6.5}$$

$$G'' = \left[\frac{\sigma_o}{\gamma_o}\right]\sin\delta \tag{6.6}$$

$$\tan\delta = \frac{G''}{G'} \tag{6.7}$$

where G′ (Pa) is the storage modulus, G″ (Pa) is the loss modulus, and tan δ is the loss tangent.

The storage modulus indicates the magnitude of the energy that is stored in the material or recoverable per cycle of deformation. The loss modulus expresses the energy that is lost as viscous dissipation per cycle of deformation. Therefore, for a perfect elastic material, all the energy is stored such that G′ is zero and the stress and strain will be in phase (Figure 6.1). On the other hand, for a liquid with no elastic property, all the energy is dissipated as heat; that is, G′ is zero and the stress and strain is out of phase by 90° (Figure 6.1). For a specific food, magnitudes of G′ and G″ are influenced by frequency, temperature, and strain. For strain values within the linear range of deformation, G′ and G″ are independent of strain. These viscoelastic functions have been found to play important roles in the rheology of polysaccharides.

Temperature sweep is well suited for studying gelatinization of starch dispersion during heating. Thus, G′ and G″ can be determined as functions of temperature at fixed angular frequency within the linear viscoelastic range. Burke and Johnson (2001) used a temperature sweep test to study the effects of a variety of wheat, growth site, and weather on gelatinization temperature of a wheat dough system. A rheometer with a 40-mm plate and plate geometry was used. They chose the temperature corresponding to G′ = 25000 Pa as a marker of the gelatinization event. The stiffness of the dough after gelatinization is mostly due to the strength of the starch phase. Brouillet-Fourmann et al. (2002) also studied gelatinization and gelation of cornstarch at various total water contents (35%, 40%, and 50% on a wet basis) by using a dynamic mechanical spectroscopy equipped with 25-mm parallel plate geometry.

Xue and Ngadi (2006) reported that storage modulus G' and loss modulus G'' followed a similar trend for batter samples formulated with different combinations of wheat, rice, and corn flours. There was no appreciable elastic response (G') below 50°C, but between 58 and 68°C, there was a rapid increase in G'. This indicates an increase in elastic properties (Figure 6.6). G' max and G'' max were higher in batter with corn and rice than in wheat alone. This was adduced to the fact that gluten gelatinizes more quickly than do corn (with high starch and fat content) and rice (with high moisture content). Figure 6.7 and Figure 6.8 show the effect of salt on storage

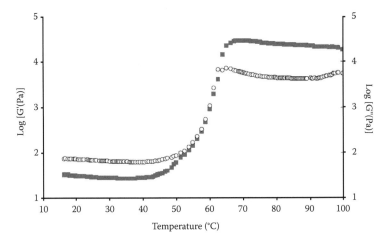

FIGURE 6.6 Dynamic oscillatory rheogram for wheat-based batter. The solid square refers to storage modulus (G') and the open circle refers to loss modulus (G''). *Source*: Xue, J. and Ngadi, M. 2006. Rheological properties of batter systems formulated using different flour combinations. *J. Food Eng.* 77(2): 334–341.

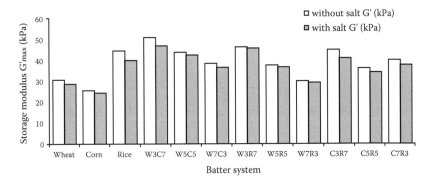

FIGURE 6.7 Effect of salt on storage modulus of batters formulated with blends of wheat and corn flour. *Source*: Xue, J. and Ngadi, M. 2006. Rheological properties of batter systems formulated using different flour combinations. *J. Food Eng.* 77(2): 334–341.

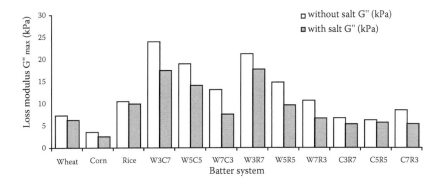

FIGURE 6.8 Effect of salt on loss modulus of batters formed using combined wheat and corn flours. *Source*: Xue, J. and Ngadi, M. 2006. Rheological properties of batter systems formulated using different flour combinations. *J. Food Eng.* 77(2): 334–341.

and loss modules of the batter systems, respectively. Salt lowered the storage module G' and the loss module G'' for the batter samples.

This was attributed to the influence of salt on the dynamic properties, which apparently resulted in a distinctly more viscous behavior. Llorca, Hernando, Perez-Munuera, Fiszman, and Lluch (2001) explained that salt affected the solubilization of gluten proteins in wheat flour, resulting in decreasing consistency coefficients. Similar results were also described by Salvador, Sanz, and Fiszman (2003). The results indicate that products containing salt would be less elastic than those samples that do not contain salt.

6.2   POROSITY

Porosity, also referred to as voidage, is the fraction of the bulk volume of a porous medium that is occupied by pores or void spaces. It is a significant factor affecting oil uptake during deep-fat frying. Fried batter porosity is significantly influenced by the formulation and composition of the batter and that of the product it is used on as coating, the frying conditions, and the pre-frying treatments (Adedeji and Ngadi, 2007, 2008a, 2008b). Initial porosity represents the void fraction of the product and reflects the volume available for oil uptake. However, an increase in porosity causes a higher oil uptake, simultaneously reducing porosity, making porosity a difficult parameter to evaluate accurately during deep-fat frying. This is because fat intrudes into the pore spaces evacuated by moisture, thus modifying the actual values of porosity. To clarify this possible confusion, Pinthus, Weinberg, and Saguy (1995) introduced a new variable called "net porosity" to account for the change in porosity due to oil uptake. A linear relationship was established between net porosity and oil uptake (Pinthus et al., 1995). Dogan, Sahin, and Sumnu (2005) measured the influence of soy and rice flour on porosity of fried chicken nuggets. Samples with soy flour were reported to have higher porosity at the later stage of frying. This was attributed to the higher viscosity of soy flour samples when compared to rice flour or control samples. The formulation of batter and breading was found to be an effective means of controlling porosity during crust formation. For example, oil absorption is affected by breading porosity. Porous breading particles pick up oil and release moisture during cooking more quickly than do dense breadings. The effective heat transfer rate is also higher in porous materials, partly due to the greater thermal conductivity of oil in the breading.

Adedeji and Ngadi (2008a) studied the effect of batter formulation on microstructural properties of fried chicken nugget coatings. The authors found that addition of carboxymethyl cellulose up to 1% into batter of 100% wheat flour reduced batter porosity substantially and concurrently fat content. This was attributed to the water holding capacity of gums, in general (Mallikarjunan, Chinnan, Balasubramaniam, and Phillips, 1997; Ford, 1999), and to the limited pores formed because less water was vaporized during frying.

Pore size and surface area are among the parameters of pores in foods. Data on these parameters for batters are very scarce. Pore size may be defined as a portion of a pore space bounded by solid surfaces. However, in reality, the actual delineation of pore size can be challenging. The surface area of a pore is the interstitial surface area of the voids and is defined as either per unit mass or per unit bulk volume of the porous material. It plays an important role in a variety of applications involving porous foods. Specific surface area determines the adsorption capacity of various food products. Thus, it influences the effective fluid conductivity, permeability, hydration, and re-hydration ability of various products. Fortuna et al. (2000) examined pore characteristics of potato, wheat, and corn starches. The authors observed that specific pore surface area of the starch granules was related to their pasting temperature and maximum viscosity. Thus, it appears that the size of pores and their specific surface areas can significantly influence their thermal and physiochemical properties. This finding may have interesting implications for breaded fried foods that typically use starch-based coatings.

Characterization of microscopic pore structure in foods is extremely challenging due to the peculiar geometry of pore structures and the complexity of food products. Mercury porosimetry, gas sorption, and microscopic imaging are typically used to determine microscopic pore structure parameters. Microscopic pore structure parameters include pore size (diameter, volume), pore size distribution, and the volume distribution function. The pore size of samples and their distributions can be measured using a mercury porosimeter. Assuming that the pores are cylindrical in shape, pore sizes could be calculated using the Washburn equation. Pore size distribution (PSD) is the table or graph of relative frequencies of pore sizes and is usually determined by fitting an experimental data to Washburn's model. PSD is commonly determined with a mercury porosimeter (Kassama, Hgadi, and Raghavan, 2003). The volume of mercury penetrating a sample is measured over a range of pressures. The pore size is calculated based on the capillary pressure obtained from the Laplace equation and using the bundle of capillary tubes (cylindrical) model of pore structure. Frying adjusts pore sizes and pore size distribution (Ngadi, Kassama, and Raghavan, 2001; Kassama et al., 2003). The nature of the influence of frying on pore size and pore size distribution is yet to be clearly elucidated. Considerable progress has been made in the last decade on studies related to microstructure of foods and the effect of processing parameters. Microstructural evaluations of fried food products tend to be largely quantitative due to the complexity of food structures. These make it difficult to evaluate the influence of pertinent process and product parameters quantitatively. More studies are required to advance knowledge in this regard.

6.3 ADHESION

Adhesion is the chemical and physical binding of a coating with itself and the food substrate. A large variety of proteins, starches, and gums aid adhesion. Adhesion is an important property of a batter and is affected by the batter formulation, food surface properties, and the cooking methods (Suderman and Cunningham, 1980). The most significant factors affecting batter adhesion are the batter viscosity and the protein content of the flour (Hsia et al., 1992). However, food formulators can choose from a wide variety of proteins, including fresh or powdered egg whites, whey powders, soy flours, and non-fat dry milk. Gelatin is quite effective at improving adhesion.

Adding other carbohydrates, especially gums, can also improve adhesion and help bind pieces together in batters used on products such as onion rings. When added to the batter, they also improve bonding. Cold-swelling thickeners, such as guar and carob bean gum can increase batter viscosity. Methylcellulose promotes adhesion between the coating material and the substrate as the result of interfacial activity and its thermal-gelling binding properties. The customer can select an appropriate grade of methylcellulose to obtain the desired viscosity for batter-solids stabilization and coating adhesion. In certain applications, some blistering or fat holdout also may be desired.

This raw material produces a significant effect on the adhesion of the breading or batter. For example, a thick layer of ice on frozen substrates results in poor adhesion. The same problem occurs if the moisture level of products such as fish is too high. This may result in voids, areas bare of any coating. In coatings with poor adhesion, the steam generated during frying causes coating "pillowing" — the formation of air pockets — or "blow-off" — the loss of coating in the frying medium.

The mixing and coating process itself can affect adhesion. If the mixing procedure or formulation does not allow the starches and gums to hydrate fully, they may not function properly.

6.4 PARTICLE SIZE

Another important property of a breading and batter system is particle size distribution. It has tremendous influence on other properties of the batter such as viscosity, porosity, and adhesion (Mukprasirt, 2000). Industrially available breadings are broadly grouped into three size categories. These are small (for breadings that have particle sizes less than the U.S. No. 60 mesh, mesh opening 250 μm), medium (particle size between U.S. No. 20 and U.S. No. 60), and large (particle size greater than U.S. No. 20 mesh, mesh opening 850 μm). In practice, breading used for coating products is a mixture of the three different particle sizes. Maskat and Kerr studied the effect of particle size on batter coating characteristics. The authors used cracker-meal breading that was grouped into the three size categories. Batters were used to coat chicken breasts. The coatings that were made with larger size particles absorbed less oil during frying but appeared rough on the surface with visible granules. In another study, Maskat and Kerr (2004) examined the effect of breading particle size on coating adhesion to the core of a coated product. Coating adhesion was estimated by shaking pre-weighed fried samples on an orbital shaker. The amount of coating loss was determined by weight difference. The authors showed that there was a difference in the adhesion of breadings of different sizes only when the shaking speed was greater than 150 rpm, at which point coating loss increased with increasing particle size. Thus, coating loss depends on how the products are handled. Scanning electron mnicroscopic images of the coated products showed that coatings made from small particle size breading had a continuous matrix between breading and batter, whereas breading particles were more visible and less of the continuous matrix in the coatings formed from medium particle size breading. For coatings formed from large particle size breading, the continuous matrix did not cover the breading particles. Instead, it formed structures connecting the large breading particles.

6.5 REFERENCES

Adedeji, A.A. and Ngadi, M.O. 2007. Pore characteristics of chicken nuggets breading coating. A paper presented at Northeast Agricultural and Biological Engineering Conference, July 27–30, Aberdeen.

Adedeji, A.A. and Ngadi, M.O. 2008a. Microstructural characterization of deep-fried breaded products using x-ray micro-computed tomography. Paper No. M06. In 10th ICEF, International Congress of Engineering and Food. Viña del Mar, Chile.

Adedeji, A.A. and Ngadi, M.O. 2008b. The use of X-ray micro-CT for characterization of microstructural properties of deep-fat fried breaded chicken nuggets. Paper No. 084618. In Annual International Meeting of American Society of Agricultural and Biological Engineers. Providence, RI: American Society of Agricultural and Biological Engineers.

Ang, J. 1993. Reduction of fat in fried batter coatings with powdered cellulose. *J. Am. Oil Chem. Soc.* 70(6): 619–622.

Baixauli, R., Sanz, T., Salvador, A., and Fiszman, S.M. 2003. Effect of the addition of dextrin or dried egg on the rheological and textural properties of batters for fried foods. *Food Hydrocolloids* 17(3): 305–310.

Brouillet-Fourmann, S., Carrot, C., Lacabanne, C., Mignard, N., and Samouillan, V. 2002. Evolution of interactions between water and native cornstarch as a function of moisture content. *J.. Appl. Polymer Sci.* 86(11): 2860–2865.

Corey, M.L., Gerdes, D.L., and Grodner, R.M. 1987. Influence of frozen storage and phosphate predips on coating adhesion in breaded fish portions. *J. Food Sci.* 52(2): 297–299.

Cunningham, F.E. and Tiede, L.M. 1981. Influence of batter viscosity on breading of chicken drumsticks. *J. Food Sci.* 46(6): 1950–1950.

Dogan, F.S., Sahin, S., and Sumnu, G. 2005. Effects of soy and rice flour addition on batter rheology and quality of deep-fat fried chicken nuggets. *J. Food Eng.* 71(1): 127–132.

Fiszman, S.M., and Salvador, A. 2003. Recent developments in coating batters. *Trends in Food Sci. Technol.* 14(10): 399–407.

Ford, J.L. 1999. Thermal analysis of hydroxypropylmethyl cellulose and methylcellulose: Powders, gels and matrix tablets. *Intl. J. Pharmaceutics* 179(2): 209–228.

Fortuna, T., Januszewska, R., Juszczak, L., Kielski, A., and Palasinski, M. 2000. The influence of starch pore characteristics on pasting behaviour. *Intl. J. Food Sci. Technol.* 35(3): 285–291.

Hsia, H.Y., Smith, D.M., and Steffe, J.F. 1992. Rheological properties and adhesion characteristics of flour-based batters for chicken nuggets as affected by three hydrocolloids. *J. Food Sci.* 57(1): 16–18.

Kassama, L.S., Ngadi, M.O., and Raghavan, G.S.V. 2003. Structural and instrumental textural properties of meat patties containing soy protein. *Intl. J. Food Properties* 6(3): 519–529.

Llorca, E., Hernando, I., Perez-Munuera, I., Fiszman, S.M., and Lluch, M.A. 2001. Effect of frying on the microstructure of frozen battered squid rings. *Eur. Food Res. Technol.* 213: 444–455.

Loewe, R. 1990. Ingredient selection for batter systems. In: *Batters and breadings in food processing*, Kulp, K. and Loewe, R. (Eds.). St. Paul, MN: American Association of Cereal Chemists, pp. 11–28.

Loewe, R. 1993. Role of ingredients in batter systems. *Cereal Foods World* 38(9): 673–677.

Mallikarjunan, P., Chinnan, M.S., Balasubramaniam, V.M., and Phillips, R.D. 1997. Edible coatings for deep-fat frying of starchy products. *Lebensmittel-Wissenschaft und-Technologie* 30(7): 709–714.

Meyers, M.A. and Conklin, J.R. 1990. Method of inhibiting oil adsorption in coated fried foods using hydroxypropyl methyl cellulose. U.S. Patent No. 4,900,573.

Mukprasirt, A., Herald, T.J., Boyle, D.L., and Boyle, E.A. 2001. Physicochemical and microbiological properties of selected rice flour-based batters for fried chicken drumsticks. *Poult. Sci.* 80(7): 988–996.

Mukprasirt, A., Herald, T.J., and Flores, R.A. 2000. Rheological characterization of rice flour-based batters. *J. Food Sci.* 65(7): 1194–1199.

Naruenartwongsakul, S., Chinnan, M.S., Bhumiratana, S., and Yoovidhya, T. 2004. Pasting characteristics of wheat flour-based batters containing cellulose ethers. *Lebensm.-Wiss. u.-Technol.* 37: 489–495.

Ngadi, M.O., Kassama, L.S., and Raghavan, G.S.V. 2001. Porosity and pore size distribution in cooked meat patties containing soy protein. *Can. Biosyst. Eng.* 43: 317–324.

Olewnik, M. and Kulp, K. 1990. Factors affecting characteristics of wheat flour in batter. In: *Batters and breading in food processing*, Olewnik, M. and Kulp, K. (Eds.). St. Paul, MN: American Association of Cereal Chemists, pp. 106–107.

Oliver, G. and Sahi, S.S. 1995. Wafer batters: A rheological study. *J. Sci. Food Agricult.* 67(2): 221–227.

Parinyasiri, T., Chen, T.C., and Reed, R.J. 1991. Yields and breading dispersion of chicken nuggets during deep-fat frying as affected by protein content of breading flour. *J. Food Process. Preserv.* 15(5): 369–376.

Pinthus, E.J., Weinberg, P., and Saguy, I.S. 1995. Oil uptake in deep fat frying as affected by porosity. *J. Food Sci.* 60(4): 767–769.

Prakash, M., Haridasa Rao, P., Susheelamma, N.S., and Prabhakar, J.V. 1998. Rheological characteristics of native and steamed wheat flour suspensions. *J. Cereal Sci.* 28(3): 281–289.

Prakash, M. and Rajalakshmi, D. 1999. Effect of steamed wheat flour on the sensory quality of batter coated products. *J. Food Quality* 22(5): 523–533.

Prakash, M., Ravi, R., and Susheelamma, N. 2001. Rheological studies of raw and steamed wheat flour suspensions with added ingredients. *Eur. Food Res. Technol.* 213(2): 113–121.

Rao, M.A. 1999. Flow and functional models for rheological properties of fluid foods. In: *Rheology of fluid and semisolid foods. Principles and applications*, Barbosa-Cánovas, G.V. (Ed.). Gaithersburg, MD: Aspen Publishers Inc., pp. 25–59.

Rasper, V. (1980). Theoretical aspects of amylographology. In: *The amylograph handbook,* Shuey, W.C. and Tipples, K.H. (Eds.). St. Paul, MN: The American Association of Cereal Chemists, pp. 1–6.

Salvador, A., Sanz, T., and Fiszman, S. 2002. Effect of corn flour, salt, and leavening on the texture of fried, battered squid rings. *J. Food Sci.* 67(2): 730–733.

Salvador, A., Sanz, T., and Fiszman, S.M. (2003). Rheological properties of batters for coating products — effect of addition of corn flour and salt. *Food Sci. Technol. Int.* 9(1): 23–27.

Sasiela, R.J. 1990. Troubleshooting techniques for batter and breading systems. In: *Batters and breading in food processing*, Kulp, K. and Loewe, R. (Eds.). St. Paul, MN: American Association of Cereal Chemists.

Shelke, K., Hoseney, R.C., Faubion, J.M., and Curran, S.P. 1992. Age-related changes in the properties of batters made from flour milled from freshly harvested soft wheat. *Cereal Chem.* 69(2): 145–147.

Shih, F. and Daigle, K. 1999. Oil uptake properties of fried batters from rice flour. *J. Agri. Food Chem.* 47(4): 1611–1615.

Suderman, D.R. 1990. Effective use of flavorings and seasonings in batter and breading systems. In: *Batters and breadings in food processing,* Kulp, K. and Loewe, R. (Eds.). St. Paul, MN: American Society of Cereal Chemists, p. 73.

Suderman, D.R. and Cunningham, F.E. 1980. Factors affecting adhesion of coating to poultry skin, effect of age, method of chilling, and scald temperature on poultry skin microstructure. *J. Food Sci.* 45(3): 444–449.

Xue, J. and Ngadi, M. 2006. Rheological properties of batter systems formulated using different flour combinations. *J. Food Eng.* 77(2): 334–341.

Xue, J. and Ngadi, M. 2007. Rheological properties of batter systems containing different combinations of flours and hydrocolloids. *J. Sci. Food Agri.* 87(7): 1292–1300.

7 Batter and Breading Ingredients Selection

7.1 FUNCTIONALITY OF INGREDIENTS

Ingredients serve numerous important functions in batter and breading systems to give their unique characteristics and functionalities to the coating. The selection of appropriate ingredients directly influences the quality of the finished product.

7.1.1 FLOUR

Flours, as generally understood, include the finely ground endosperm of wheat. In the literature of batters and breadings, however, flour appears to refer to finely ground starchy material from any of several sources. Flour functionality in batter and breading systems depends in large part on the two major constituents of all flours: starch and protein.

Flour is the key ingredient in batter and breading systems. Flours provide viscosity and may promote adhesion through the formation of gluten, which provides structure and texture and can act as a barrier to fat absorption. Flour contains some reducing sugars that caramelize during frying, contributing to the color and flavor of the coating (Mohamed, Norhasimah, and Mansoor, 1998). Because flour is the main component of most breadings, it can be used as is or first be baked into a crumb. The porosity of the resulting products affects the oil absorption; the more porous the material, the more oil is absorbed.

The most typical flour used in batters and breadings is wheat flour. A dry batter mix formula generally contains 80 to 90% flour, which is primarily wheat flour. When wheat gluten is added to a batter mix, its film-forming properties reduce moisture loss and produce crisp, appetizing surfaces. Additionally, pre-dusting food with wheat gluten significantly improves adhesion and enhances the appearance (Magnuson, 1985). Wheat flour with higher protein levels will increase batter viscosity and produce darker and crisper fried food. Most manufacturers are not looking for their flour to provide protein adhesion, so they often choose flour with a protein level in the range of 10.5 to 11.2%.

Although wheat flour is the primary flour used, the use of corn and rice flours is increasing. In addition to rice and corn, soy, malted barley, and potato flours also may be used.In a study conducted by Flores, Herald, and Mukprasit (2000), it was shown that RFBB could be substituted for wheat flour-based batters (WFBB). Proteins and starches in rice flour are chemically different from those in wheat flours and retain weaker bonds with oil (Shih and Daigle, 1999). They reported that rice flour resisted oil absorption better, but was less effective than wheat flour as a thickening agent. Their results showed a 69% oil reduction with rice flour batter on shrimp products. Due to the weaker bonds with oil, RFBB may produce a healthier product because there is less absorption of oil by the food material. This is consistent with the findings in a study conducted by Mukprasirt, Herald, Boyle, and Boyle (2001) in which the use of RFBB as an alternative to the traditional WFBB was investigated. It was found that all RFBB exhibited lower oil absorption than WFBB. Additionally, RFBB had a higher breading pickup than WFBB. Thus, rice flour shows its potential to serve as an alternative to wheat flour in battered and breaded foods. RFBB might be a commercially feasible new product in the food industry. A high ratio of rice flour provided a roughness to the crust, however; RFBB forms thin slurries and requires additives to develop viscosity and other desirable batter properties. A good strategy is to use rice-based thickening agents as additives. For example, gelatinized long grain rice flour and phosphorylated long grain rice starch ester can be effective in enhancing the batter viscosity and the oil-lowering properties of rice flour batters. Mukprasirt, Herald, Boyle, and Rausch (2000) also studied the effects of ingredients used in RFBBs on the adhesion characteristics for deep-fat fried chicken drumsticks. They found that batter formulated with a 50:50 mixture of rice and corn flours adhered better to drumsticks than did batter with other rice flour ratios (30:70, 70:30, rice flour:corn flour). As the rice flour ratio increased from 50 to 70%, the binding force decreased. They suggested combining rice flour with other ingredients. For example, methylcellulose, oxidized starch, and xanthan gum increased the amount of batter pickup before frying by increasing viscosity, and achieved finished products with lower fat content. As a result, foods coated with corn flour-based batters are generally more crisp (Loewe, 1993).

7.1.2 STARCH

Traditionally, starches have been used for adhesion in batters. Starch consists of two main polysaccharides: amylose and amylopectin. Both polysaccharides are based on chains of 1d4 linked α-D-glucose. Amylose is essentially linear, whereas amylopectin is highly branched containing, on average, one 1d6 branch point for every 20 to 25 straight chain residues (Hoseney, 1994). Starches from different plant sources consist of different ratios of amylose to amylopectin (Pomeranz, 1991; Park, Ibáñez, and Shoemaker, 2007). In normal cereals, the amylose content is approximately 23 ± 3% (Hoseney, 1994). In some applications, the ratio of amylopectin to amylose has a profound effect on the functionality of the starch. This ratio is quite variable, ranging from 99% amylopectin in waxy rice, waxy corn, and waxy sorghum to 75% amylose in high amylose corn. Starches containing higher quantities of amylose typically produce batters with enhanced textural qualities. High amylose (70%) starches are especially useful in batter coatings that require a continuous membrane (Loewe, 1993). Starches with higher amylose content are generally selected for better film-forming properties. They produce a crisper and stronger film, which stays intact through frying. Therefore, when batters containing these types of starches are applied to foods, oil uptake is significantly reduced. High amylose starches are also especially useful in batter coatings requiring microwave heating. Oxidized starches are used for their basic adhesion and coating, while high amylose starches help reduce fat pickup. Batters typically contain starch levels of 5 to 30% of the dry mixture (Mukprasirt et al., 2001).

Starch granules occur in varying sizes and the range and distribution of sizes depend on the source of the starch. Within intact cereal grains, the starch granules are embedded in a protein matrix. The degree of binding between the protein and starch and the milling process used to separate them affects particle size distribution and the degree of starch damaged in the resulting flour, which in turn affects the functional properties of the flour. Flour from hard wheat tends to contain a greater proportion of damaged starch than does flour from soft wheat (Loewe, 1993). Damaged starch granules absorb higher amounts of water than do intact granules (which absorb water approximately 30% of their weight). Therefore, more water would be needed for a given batter viscosity when flours with higher levels of damaged starch are used.

In water at room temperature, the native starch granules of flour tend to settle quickly out of suspension. When heated in the presence of water, starch undergoes a process known as gelatinization. Initially, the process involves swelling of the granules as they absorb water into portions of their structure. In the original granule structure, a considerable portion of the starch exists in a crystalline form, which is impervious to water at moderate temperatures. When heated, these areas are broken down and hydrophilic portions of the starch molecule are exposed to water, resulting in swelling of the sample. Additionally, some material from within the starch granule is exuded into the solution where it acts to further increase the viscosity of the solution. The gelatinization temperature of starches ranges from 52 to 63°C for wheat starch and 69 to 75°C for sorghum starch (Loewe, 1993). When cooled, gelatinized starch molecules re-aggregrate in a process known as retrogradation, leading to a gel formation that gradually increases in firmness with time and lower temperature. The rate and degree of retrogradation are much greater for linear amylose than for branched amylopectin.

Gelatinized starch, serving as the major component of flour or as a separate purified ingredient in a batter or breading formula, becomes the major framework of the final product coating. In fact, acceptable batters and breadings are possible with just water and flour, although they tend to be somewhat flat, smooth, and uninteresting in appearance as well as lacking in flavor. To best achieve an even base coating, it is obviously important to distribute the starch portion of the batter or breading evenly about the product, ensuring that upon gelatinization an even coating of starch gel is formed, which will completely envelop the product.

Pure starch is usually produced through a wet milling process, which results in little or no starch damage. Pure starch does not stay suspended in water nor does it carry sufficient water by itself to produce complete gelatinization in a conventional frying system. Pure, unmodified starches are generally used in conjunction with flour in batter and breading formulations. Addition of isolated starch to a flour containing batter/breading formulation will provide an effect comparable to reducing the protein and damaged starch levels of the flour. Modified starches with a wide range of hydration, adhesive, and film-forming characteristics are also available.

Use of either unmodified or modified starches in batter/breading mixes allows the formulator to work outside the range of properties that may be available through the selection of flour type. In mixes containing both flour and starch (modified and unmodified), variations in the ratio of flour to starch or changes in starch type (degree of gelatinization) may allow for adjustment to variations in the flour, which are likely to occur from time to time. Kuntz (1995) reported that flour-based batter absorbed more water and settled more quickly than did batter that contained starches.

7.1.3 PROTEIN

Adding protein, which might be added at a level of 10 to 15%, helps the structure or changes the texture of the final coated product. Research shows that products with higher protein contents are generally more effective as binding agents. Protein has been used to improve the water absorption capacity of flour, which in turn increases the viscosity of the system. It is also used to strengthen the structure and texture, retard moisture loss, and enhance crust color and flavor development. The level of flour protein used has a major effect on batter pickup, ranging from 11 to 28% when measured at equal water–solids ratios (Loewe, 1993). In general, a higher level of protein increased crispness of the fried product and produced a darker color. As the protein level increased, there was a gradual increase in roughness of texture and brittleness of the fried coating. The pancake-like inner structure was no longer present in the high-protein (12.1%) flour coatings.

Gelatinized starch, along with flour protein, forms the structure of the final batter-coated product. Wheat flour, when included in a batter system that undergoes any significant amount of mixing, will show a gradual increase in viscosity due to the development of gluten protein. Gluten is made up of two groups of large protein molecules which, when hydrated, interact to form a cohesive matrix. The gluten matrix may support a structure varying from as firm as bread dough to a low viscosity and very flowy batter depending on the quantity and, to some extent, the quality of the gluten and the relative amount of water available. Hard wheat flours, due to their higher protein content, require more water than do soft wheat flours to yield comparable viscosities when used in a batter. This results from the efficient water-binding capacity of the gluten protein fraction.

In many batter and breading formulas, no particular type of flour is specified. Formulas where a flour type is mentioned generally call for all-purpose flours. Viscosity is essential to a hydrated batter because many ingredients are insoluble at ambient or refrigerated temperatures. A more viscous batter brings about the suspension of these ingredients, preventing undesirable stratification. The absorptive capacity of flour protein aids in maintaining uniform dispersion of such ingredients for optimum performance.

In puff/tempura batters, gluten proteins help gas retention during leavening. The resulting formation of an aerated, porous, cooked batter is essential for proper texture and crispness.

7.1.4 CHEMICAL LEAVENING

Leavening agents are usually used in tempura batters. The primary function of leavening agents is to provide gas for aeration and expansion of batter during mixing and frying (Chung and Edging, 2000). There are essentially two components in a chemical leavening system: bicarbonate, which supplies carbon dioxide, and an acid that controls the release of carbon dioxide from bicarbonate in the presence of water (Chung and Edging, 2000). Some batters use a mixture of acids, one to provide some gas release at room temperature and another that reacts at a higher temperature to provide leavening action during the actual frying process (Davis, 1983). The gas-release characteristic of a specific leavening affects the texture. If the release is too early, the product texture will be coarse and the coating will absorb excess oil. Additional leavening may be able to change the color and texture of a fried product. For example, a corn dog coating needs a leavening system that releases gas very rapidly so that the coating can expand very quickly. The batter becomes more brittle as the amount of leavening is increased.

The leavening system can be tailored for a specific application by varying the type of leavening acids incorporated. The amount of gas generated and its rate of production determine the effects in the batter. Dubois (1981) mentioned that typical leavenings used in batters include sodium acid pyrophosphate (SAPP), sodium aluminum phosphate (SALP), and combinations of SALP and monocalcium phosphate (MCP).

7.1.5 SHORTENING AND OIL

The shortening system plays a key role in the mouth feel or eating quality of battered and breaded food (Landes and Blackshear, 1971). Shortening and oil have specific functions in coatings such as being carriers of fat-soluble vitamins and contributing to food flavor and palatability as well as to the feeling of satiety after eating. Other fatty materials with potential use in batters and breadings include emulsifiers and staling inhibitors (Fennema, 1976). The melting point and solid content are functions of the source of the oil selected for the frying shortening.

7.1.6 EGGS

Eggs are used widely, both as a batter ingredient and as pre-dips. Egg contains albumin, a heat-sensitive protein that is useful in binding the breading/batter to the substrate. Lecithin in the yolk portion may act as an emulsifier and contribute to batter stability. The white's protein improves adhesion, while the yolk's phospholipids provide increased emulsification. Egg white may create some microbiological issues especially if the product is stored on a line at a room temperature of 70° to 90°F (Gerdes, 2001). The addition of egg to a batter tends to darken the final product (Loewe, 1993).

7.1.7   MILK AND WHEY

Dairy ingredients contribute flavor, adhesion, and color. Added as liquids or dry powders, these ingredients provide lactose, a reducing sugar that is involved in browning reactions, and protein, which provides structure and participates in non-enzymatic browning. Typical pre-dips include milk, evaporated milk, and buttermilk. Some producers select buttermilk for its unique flavor profile in many coating food recipes, including rock shrimp cones, crispy fried chicken, and fried okra.

7.1.8   FLAVORING AND SEASONING

One way to improve the value-added perception of battered and breaded products is to incorporate additional flavorings and seasonings. Donahoo (1970) stated that the seasoning level varies considerably although the average is 3 to 5% of a batter mix. Spices and herbs are dry ground plant materials that possess a characteristic taste and contain many aromatic and flavor constituents. They are usually used at a rate of 0.5 to 1.0% in finished food products. At the same time, some spices may contribute to development of specks and colors that are unacceptable (i.e., paprika). It also takes time to reach flavor equilibrium with their medium (Suderman 1993). Essential oils offer many advantages including high flavor concentration with no off-colors or specks, instant flavor equilibration, and easy blending and quality control. They are used in coated food products at a level of 0.01 to 0.10% of the finished weight.

Flavor components are released from the coated food (substrate, pre-dust, and batter-breading complex). All flavor components have a volatility spectrum, and it is conceivable that some flavor components are steam stripped from the food as water converts to steam and exits the food surface. Flavor components are also imparted to the coated food product by the frying oil or heat stable flavors contained in the frying oil. In the commercial batter and breading mix, garlic is one of the popular ingredients that improve the flavor of the product. From a food safety standpoint, it should be kept in mind that spices are occasional sources of microbial contamination.

Batter coatings can easily have seasoning blends incorporated into the dry batter mix, either as an internal flavor system or as an external flavor system. Internal flavor systems incorporate flavorants during the breading making process. These internal flavorants can be added directly to the breading mix. In external flavor systems, flavorants are sprinkled onto a "wetted" breading surface via oil spray, for example.

7.1.9   SALT AND SUGAR

Salt and sugar are normally added as flavorants. Although each can compete for water and slow the rate of protein hydration in limited-water systems, this does not appear to be a problem in conventional batter formulas.

7.1.10   HYDROCOLLOIDS

Hydrocolloids are a group of polysaccharides and proteins that perform functions in food materials such as thickening and gelling aqueous solutions (Phillips and Williams, 2000). Hydrocolloids have a significant impact on the textural properties of batters and the amount of oil uptake and batter pickup (Mellma, 2003; Sanz et al., 2004). In a study conducted by Hsia, Smith, and Steffe (1992), it was shown that as hydrocolloid concentration increased, coating pickup also increased. Food gums such as xanthan are most often used to control viscosity and improve water-binding properties in various batter/breading systems. Xanthan gum was the first of a new generation of polysaccharides produced by biotechnology (Imeson, 1997), which produces highly viscous solutions even at low concentrations. This property is due to the unique rigid, rod-like conformations of xanthan in solution and its high molecular weight. It also imparts high pseudoplasticity, which is directly proportional to concentration and shear rate (Flores et al., 2000).

Hydrocolloids serve a wide variety of functions, ranging from stabilizing the entire system to replacing fats. In many foods, these ingredients are also useful for building and/or modifying the product's texture. Hydrocolloids are of special interest because they possess good barrier properties to oxygen, carbon dioxide, and lipids (Mallikarjunan et al., 1997; Williams and Mittal, 1999). Recently, a number of studies have reported the use of hydrocolloids to perform different functions in food. Various hydrocolloids (long-chain polymers), especially cellulose derivatives, form gels that can lead to reduction in oil absorption during frying. MC, HPMC, and other cellulose gums with film-forming properties form thermal gel structures that increase moisture retention and reduce fat pickup. This can be combined with a reduced-fat substrate to achieve fat levels meeting low or reduced-fat claims.

Cellulose gums in particular, not only help to regulate batter viscosity but also aid in reducing oil uptake and controlling moisture retention within the food material. MC and HPMC are the only gums used in food formulation that exhibit reversible thermal gelation, which causes batters to "set" temporarily during frying. As a result, they reduce batter "blow off" and pillowing and decrease residual debris in cooking oils (Flores et al., 2000). In a Dow Chemical Company study (1991), batters formulated with HPMC absorbed 26% less oil after a 2-min fry cycle than did the control batters. Furthermore, an addition of HPMC as a prehydrated gum solution to the batter resulted in an even greater oil reduction — up to 50% more in some applications. MC is a polymer that dissolves in water and facilitates the dispersion of other ingredients in the batter, thereby improving batter consistency and stability (Flores et al., 2000).

Food gums also have desirable interactions with starches. Small amounts (0.1 to 0.2%) of xanthan gum can help stabilize starch solutions during storage by preventing the starches from retrograding (Imeson, 1997). Many hydrocolloid substances known as gums have been used as ingredients in batters and serve three functions. The two primary functions are viscosity control and water-holding capacity. A third function with some gums results from their ability to participate in a gel or film formation in conjunction with other ingredients. This strengthens the coating and may provide greater flexibility and increased resistance to handling.

Gums provide some of the same advantages as modified starches in increasing viscosity and holding water (Meyers, 1990). They are effective at much lower levels than the modified starches: generally less than 2% of the formulation and frequently in the 0.5% range. Additionally, due to the low addition level, gums do not dilute the protein of the base flour.

Among the natural gums, xanthan has been particularly useful in viscosity control. In automated batter/breading systems, minor variations in viscosity may result in unacceptable changes in the quantity of batter pickup or breading retention, thus making viscosity control a primary formulation requirement. Xanthan has also proved valuable in low solid batters as an additive for maintaining a homogeneous suspension.

Cellulose has been modified in several ways to provide a series of ingredients useful in the modification of viscosity and maintenance of suspensions. These ingredients may be used in batters and breadings as their thermal gelation during frying provides a fat absorption-resistant coating. This reduces the amount of fat absorbed during frying, resulting in a less greasy product and reduced fat consumption. The considerations when choosing a gum in a batter system are the competition for water, tolerance to ionic species, batter pickup, water holding capacity, and water-to-solids ratio (Meyers, 1990).

Gums also improve batter adhesion. CMC forms films to keep the batter intact, and MC and HPMC aid cohesiveness due to their thermal gelatinizing properties during heating (Garcia et al., 2001). Other gums such as alginates, gellan gum, and carrageenan gelatinize in the presence of ionic salts. Guar and xanthan can increase batter viscosity in low-solids batters to ensure the correct application rate and help keep solids suspended. Because so many types and variations of gums and starches exist, selecting the correct ingredient for the desired result can be challenging. Gums generally are used at a level below 0.2% on a total battered-product basis (Dow Chemical, 1990). Moisture migration from the substrate into the batter is a concern for food manufacturers, especially in the seafood industry. Gums can help this by reducing moisture migration during frozen storage. Excessive use of gums can adversely affect the texture, creating a chewy rather than a crispy product, and can negatively affect the flavor.

7.2   FORMULATION OF BATTER AND BREADING SYSTEMS

Each batter or breading is uniquely formulated according to the specifications of the food product that it is designed to coat. While coating food products, batters contribute to the food's appearance, taste, color, juiciness, crispness, texture, and acceptability, but each one of these characteristics is also affected by the coating's ingredient makeup and cooking procedures. When these factors are not closely monitored, problems of adhesion and encapsulation appear and consumer acceptance declines. Proper formulation of batter and breading systems is essential in dealing with these problems.

7.2.1 BATTER SYSTEM

Batters can be used in two ways: either to force something to adhere to the food being coated or to serve as the final coating itself. Some specialty batters are used for other purposes, but these usages are not the norm. Donahoo (1970) described a dry mix, primarily starch and flour, which would generally comprise 80 to 90% of the total mixture, which was then combined with water on a 1.2 to 2.0:1 basis. He recommended that a batter be made up cool rather than at room temperature. Some commonly used batter systems for the food industry are shown in Table 7.1.

Adhesion batters need to be designed to provide correct viscosity, resulting in complete coverage. However, they also must have sufficient viscosity to hold breading in place. Starch or other water-absorbing materials can help set the batter to increase breading adhesion; however, ingredient selection is crucial. Thickening with ingredients that rapidly absorb a lot of water might cause coating to blow off as steam escapes. Tempura batters and similar coatings, such as fish-and-chip batters, require adhesive qualities because they have to stick to the substrate, but they also serve as the outer coating layer. Tempura batters share similar ingredients with adhesion batters, with one important difference — they contain leavening agents to greatly influence texture. Sodium bicarbonate, combined with an appropriate leavening agent, produces carbon dioxide when exposed to heat and moisture creating the characteristic "open" structure.

7.2.2 BREADING SYSTEM

Granulation, texture, density, moisture and oil absorption potential, browning rate, color, and flavor are all important breading characteristics that should be considered when selecting a breading for a particular application. A coarser breading creates a crisper product, so a finer breading should be used on relatively small food pieces like mushrooms. Therefore, granulation size is another consideration when

---

TABLE 7.1
Concentrations and Functionality of Ingredients Used in Batter Formulations

Ingredient	Addition Range (%)	Functionality or Effect on Quality
Wheat flour	> 40	"Body" structure and viscosity
Corn flour	> 30	Crispness and golden-brown color
Rice flour	< 5	Transparency, light texture, improved adhesion
Starches	< 5	Changes in tenderness and crunchiness, improved crispness
Leavening agents	< 5	Open the structure of batter upon cooking, release moisture without blowing off the coating porous structure
Gum	< 1	Viscosity control, solid suspension in cold batter, ability to participate in gel/film formation
Flavorings and seasonings	Open	Acceptable taste profiles

Source: Fiszman et al., 2003.

choosing crumbs (Mohamad and William, 2002). It affects the crumb's ability to absorb underlying batter moisture, changing its set-up rate.

If a product encounters only a brief waiting period between being battered and breaded and going through the fryer, it will need a quick-setting system with smaller crumbs to absorb more moisture. This relates to pickup, the amount of crumb that adheres to the substrate, which also is granulation-sensitive and generally increases in proportion to crumb size. Like absorption, even and complete distribution of crumb across a substrate's surface, called coverage, increases as the crumb gets finer. For a softer texture and paler crumb, go with finer granulations. The increased surface area and irregular shape of large and coarse granulations contribute to a crispy and crunchy texture and a golden color.

Foods that require extended fry times, such as breaded raw chicken parts, should use breading with less reducing sugar content. Breading may be flavored during manufacturing or later after further processing with seasonings or flavors applied to the exterior surface.

7.3 BREADING AND BATTER SELECTION

The type of starch used to formulate batters and breadings has a profound effect on viscosity, adhesion and uniformity of the batter layer, texture, and reconstitution stability. The choice of starch must be based on its functional properties. Starches are most often employed due to their versatile structures. Hydration of starches produces an irreversible change in the structure of the starch granule. Starch–starch interactions are replaced by starch–water interactions. As a result, the granule continues to swell until it finally ruptures causing starch polymers to be dispersed, which in turn produces a viscous solution (Phillips and Williams, 2000). These types of interactions allow starches to control texture and structure in food products. The chemical structure of starches allows them to be easily modified. Starch modifications are a means of altering the structure and affecting the hydrogen bonding in a controllable manner to enhance and extend their applications (Phillips and Williams, 2000).

7.4 REFERENCES

Bertram, A. 2001. Pump up the amylose. *Food Process.* 19.

Chung, F.H.Y. and Edging, T.E. 2000. Leavening acid composition. U.S. patent No. 6080441.

Davis, A. 1983. Batter and breading ingredients. In: *Batter and breading Technology,* Suderman, D.R. and Cunningham, F.E. (Eds.). Westport, CT: AVI Publishing Company.

Donahoo, P. 1970. Choosing the right batter and breading. In Proceedings of the Annual Poultry and Egg Further Processing Conference 7th, 1970.

Dow Chemical. 1990. A food technologists guide to Methocel premium food gums. Brochure 194-1138-0396 AMS. Midland, MI: The Dow Chemical Co.

Dow Chemical. 1991. Fried foods stabilizer keeps moisture in, fat out. *Prepared Foods.* 160(3): 61.

Dubois, D.K. 1981. Chemical leavening technology. American Institute of Baking, Manhanttan, KS, U.S. Bulletin 3(9).

Fennema, O.R. 1976. Principles of food science. Part 1. Food chemistry. New York: Marcel Dekker.

Flores, R.A., Herald, T.J., and Mukprasit, A. 2000. Rheological characterization of rice flour based batters. *J. Food Sci.* 65(7): 1194–1199.

Garcia, M.A., Ferrero, C., Bertola, N., Martino, M., and Zaritzky, N. 2001. Effectiveness of edible coatings from coatings from cellulose derivatives to reduce fat absorption in deep fat frying. Available at: http://ift.conex.com/ift/2001/techprogram/paper_7480.htm. Accessed June 26, 2008.

Gerdes, S. 2001. Batters and breadings liven tastes. *Food Product Design.* http://www.food-productdesign.com/articles/463/463_1201de.html. Accessed June 26, 2008.

Hoseney, R.C. 1994. Minor constituents of cereals. In: *Principles of cereal science and technology*, 2nd ed. St. Paul, MN: American Association of Cereal Chemists, Inc.

Hsia, H.Y., Smith, D.M., and Steffe, J.F. 1992. Rheological properties and adhesion characterstics of flour-based batters for chicken nuggets as affected by three hydrocolloids. *J. Food Sci.* 57(1): 16–18, 24.

Imeson, A. 1997. *Thickening and gelling agents for food.* Glasgow: Blackie Academic & Professional Publishers.

Kuntz, L.A. 1995. Building better fried foods. *Food Prod. Design* 5: 129–146.

Landes, D.R. and Blackshear, C.D. 1971. The effect of different cooking oil and color of fired chicken breading material. *J. Poultry Sci.* 50: 894–898.

Loewe, R. 1993. Role of ingredients in batter systems. *Cereal Foods World* 38(9): 673–677.

Magnuson, K.M. 1985. Uses and functionality of vital wheat gluten. *Cereal Foods World.* 30(2): 179–180

Mallikarjunan, P., Chinnan, M.S., Balasubramaniam,V.M., and Phillip, R.D. 1997. Edible coating for deep-fat frying of starchy products. *Lebensmittel-Wissenschaft und-Technologie* 30: 709–714.

Mellema, M. 2003. Mechanism and reduction of fat uptake in deep-fat fried foods. *Trends Food Sci. Technol.* 1: 364–373.

Meyers, A.M. 1990. Functionality of hydrocolloids in batter coating systems. In: *Batters and breadings in food processing.* Kulp, K. and Loswe, R. (Eds.). St. Paul, MN: American Association of Cereal Chemists.

Mohamad, Y.M. and William, L.K. 2002. Coating characteristic of fried chicken breasts prepared with different particle size breading. *J. Food Process. Preserv.* 26: 27–38.

Mohamed, S., Norhasimah, A.H., and Mansoor, A.H. 1998. Food components affecting the oil absorption and crispness of fried batter. *J. Sci. Food Agricul.* 78: 39–45.

Mukprasirt, A., Herald, T.J., Boyle, D.L., and Rausch, K.D. 2000. Adhesion of rice flour-based batter to chicken drumsticks evaluated by laser scanning confocal microscopy and texture analysis. *Poultry Sci.* 79: 1356–1363.

Mukprasirt, A., Herald, T.J., Boyle, D.L., and Boyle, E.A. 2001. Physicochemical and microbiological properties of selected rice flour-based batters for fried chicken drumsticks. *Poultry Sci.* 80: 988–996.

Park, I.M., Ibáñez, A.M., and Shoemaker, C.F. 2007. Rice starch molecular size and its relationship with amylose content. *Starch/Stärke* 59: 69–77.

Phillips, G.O. and Williams, P.A. 2000. *Handbook of hyrocolloids.* Boca Raton, FL: CRC Press.

Pomeranz, Y. 1991. *Functional properties of food components.* New York: Academic Press.

Sanz, T., Salvador, A., and Fiszman, S.M. 2004. Effect of concentration and temperature on properties of methylcellulose-added batters application to battered, fried seafood. *Food Hydrocolloids* 18: 127–131.

Sharon, G. 2001. Batters and breadings liven taste. Available at: http://www.foodproductdesign.com/archive/2001/1201de.html.

Shih, F. and Daigle, K. 1999. Oil uptake properties of fried batters from rice flour. *J. Agricul. Food Chem.* 47: 1611–1615.

Shinsato, E., Hippleheuser, A.L., and Van Beirendonck, K. 1999. Products for batter and coating systems. *The World of Ingredient.* January-February, 38–42.

Suderman, D.R. 1993. Selecting flavorings and seasonings for batter and breading systems. *Cereal Foods World* 38: 689–694.

Williams, R., and Mittal, G.S. 1999. Water and fat transfer properties of polysaccharide films on fried pastry mix. *Lebensmittel-Wissenschaft und-Technologie,* 32: 440–445.

8 Measuring the Quality of Breaded Fried Foods

The ultimate success for a food product is its quality perceived by the consumers at the time of consumption. One poor experience by a consumer translates to a huge loss of consumer confidence and acceptability of the product not only by that consumer alone but also by his friends. The unique characteristics of breaded fried foods offer challenges in objective quality measurements that are related to sensory qualities. Often times, the food service industry also has to characterize the changes in quality during storage under a heat lamp and large-scale industries manufacturing partially fried foods have to characterize the breading pickup, yield, and storage under frozen conditions. This chapter describes the methods that are being used by the industry for characterizing breaded fried food quality and discusses some recent developments in the rapid measurement of product quality.

8.1 OIL UPTAKE AND YIELD

The majority of breaded fried foods have a high moisture core and a crispy crust. High moisture retained in the core is beneficial to the industry in terms of economic benefits as well as making juicy products. In breaded fried foods, the breading and batter act as a barrier for moisture release to the frying oil by forming a partially impermeable crust. Thus, the cooking yield of the breaded fried foods is very much related to the amount of moisture lost to frying. This can be objectively measured through the estimation of moisture content in the product. The estimation of moisture in the core is more critical than the overall moisture of the fried product. However, the industry uses both overall moisture content and core moisture content. Often it is simply the mass changes in the product before and after frying that are measured.

The official standards for moisture measurement are available from AOAC (Association of Official Agricultural Chemists) and AACC (Associations of Cereal Chemists). Common methods include the use of oven drying (in either forced convection or natural convection modes), vacuum oven, and freeze dryers. Of all the methods, freeze-drying is costly but has the advantage of low temperature drying resulting in negligible changes to oil content in the fried foods. This in turn is a beneficial feature for further characterization of oil pickup in the fried foods.

Ramirez and Cava (2005) determined the moisture content of fried pork loin chops by drying 5 g samples at 102°C until a constant weight was reached (AOAC, 2000). Maskat, Yip, and Mahali (2005) determined the moisture content of surface layers of coated chicken breasts by oven drying at 105°C (AOAC, 1995). Dogan, Sahin, and Sumnu (2005) dried fried chicken nuggets in a forced convection oven at 105°C until constant weight was reached (AOAC, 1984).

Shyu, Hau, and Hwang (2005) dried vacuum-fried carrot chips in a vacuum oven to a constant weight at 70°C to determine moisture content (MC). Maskat and Kerr (2004) measured the weight difference of fried chicken breasts before and after drying at 100°C in a vacuum oven for 24 h (AOAC, 1980). Garayo and Moreira (2002) dried ground vacuum-fried potato chips in a forced air oven at 105°C for 24 h (AACC, 1986) to determine MC. Moyano, Rioseco, and Gonzalez (2002) dried French fries and Pedreschi and Moyano (2005) dried potato chips in a convection oven at 105°C until constant weight was reached. Holownia, Chinnan, Erickson, and Mallikarjunan (2000) determined the MC of fried marinated chicken strips using a freeze dryer. The pressure in the chamber was kept below 3.2 kPa and the temperature was 20°C. Innawong (2001) also used a freeze dryer to determine MC of chicken nuggets.

The oil pickup in the fried food has to be controlled for developing a breaded fried food with desirable quality attributes. This can be objectively evaluated by estimating the fat content in the product. During frying, the crust layer is involved in oil transfer. Therefore, it is critical to evaluate the oil content in the crust. In addition, industry uses overall fat content in the fried foods. Methods to evaluate fat content range from simple spectrophotometric evaluations to complex solvent-based extraction of oils in the foods. Traditional solvent extraction methods such as soxlet extraction are labor intensive and time consuming. To estimate the fat content in fried foods, many rapid extraction methods have been introduced: microwave assisted extraction (CEM), supercritical CO_2 extraction (ISCO), and ANKOM solvent extraction (a rapid extraction process using petroleum ether as the solvent under pressure at elevated temperatures, AOCS Official Procedure Am 5-04). In addition, rapid evaluation methods using near infrared (NIR) analyzers are becoming popular in the frying industry.

Kita and Lisinska (2005) used Soxhlet's method (AOAC, 1995) to determine fat content of French fries with a Buchi B-811 universal extraction system (Buchi Labortechnic AG, Flawil, Switzerland). Shyu et al. (2005) determined oil content (OC) of vacuum-fried carrot chips gravimetrically by Soxhlet extraction (AOAC, 1995). Dogan et al. (2005) used Soxhlet extraction with *n*-hexane for 6 h on fried chicken nuggets (AOAC, 1984). Ballard (2003) used method 991.36 (AOAC, 1995) with a Soxtec extraction unit to determine OC of chicken nuggets. Garayo and Moreira (2002) used the Soxtec System HT extraction unit (AACC, 1986) with petroleum ether to determine OC of vacuum-fried potato chips. Moyano et al. (2002) used a method by Bligh and Dyer (1959) to determine oil content of French fries. This method involves using a 1:2:0.8 (v/v/v) mixture of chloroform, methanol, and water. Pedreschi and Moyano (2005) also used this method on potato chips. Ramirez and Cava (2005) used the same method but with 1:2 chloroform/methanol mixture. Innawong (2001) used method 960.39 (AOAC, 1995) to determine OC of freeze-dried chicken nuggets. Holownia et al. (2000) used a Labconco Goldfisch Fat and Oil Extractor (Labconco Corp., Kansas City, MO) with petroleum ether as the solvent to determine OC of freeze-dried edible film-coated chicken strips.

8.2 BREADING PICKUP AND ADHESION

The breading pickup or breading adhesion is calculated in the food industry as described by Suderman and Cunningham (1983). Ten pieces of fried products are placed in a standard wire sieve (No. 4 U.S. sieve) and shaken for 1 min. The breadcrumbs that accumulate in the catch pan are weighed and the percentage of coating loss is calculated. The breading pickup is a critical factor for cost reduction in producing the product.

Maskat and Kerr (2004) shook fried chicken breasts on an orbital shaker (VMRbrand, VMR Scientific Products, West Chester, PA). Innawong (2001) measured breading adhesion using a method described by Suderman and Cunningham (1983), in which 10 pieces of the fried product (in this case, chicken nuggets) were shaken in a standard wire sieve (No. 4 U.S. sieve) for 1 min. Accumulated breadcrumbs were weighed and used to calculate percentage coating loss.

8.3 COLOR AND APPEARANCE

The first and foremost quality by which a consumer makes a decision is the visual appearance. This also affects the taste perception of fried foods. Color can be defined as the energy distribution of the light reflected by a fried food product. The reflected light is affected by the light source, the angle of viewing, and the characteristics of the background. Together, all these factors affect the color perception. Often, the color is measured using standard CIE (The Commission International de l'Eclariage) scales. In the food industry, the derived color scales such as Hunter L, a, and b or CIE L*, a*, and b* are commonly used. These parameters represent the lightness (L), the degree of redness or greenness (_a), and the degree of yellowness or blueness (_b) of the product. A value of 0 or 100 for L represents black or white, respectively. These instrumental color measurements correspond to the human perception of the color such as hue, saturation, and value (lightness).

The hue describes the visual sensation of the color. The hue represents the appearance of a given area in comparison to one or proportions of two or more of the perceived standard colors red, yellow, green, and blue. Chroma, the purity or saturation, describes the intensity of a fundamental color with respect to the amount of white light that is mixed with it. In other words, converting the coordinate system of "a" and "b" to polar coordinates of "r" and "θ" gives chroma and hue, respectively. Thus, the hue angle and chroma can be calculated as:

$$\text{Chroma or Saturation} = \left[a^2 + b^2 \right]^{0.5}$$

$$\text{Hue Angle} = \tan^{-1}\left(\frac{b}{a} \right)$$

The value or lightness is an indication of overall light reflectance of that color. Thus, the color parameter L* represents the value or color lightness.

In many instances, the differences in color parameters between two stages provide more meaningful interpretation of the processes than do absolute color values. The color differences between two stages (e.g., between raw and frozen food in a frozen stage, during storage, or in a thawed stage) can be calculated as:

$$\text{Hue Angle Difference} = \tan^{-1}\left(\frac{b}{a}\right) - \tan^{-1}\left(\frac{b_0}{a_0}\right)$$

$$\text{Saturation Difference} = \left[(a - a_0)^2 + (b - b_0)^2\right]^{0.5}$$

$$\text{Brightness Difference} = \text{abs}(L - L_0)$$

$$\text{Total Color Difference} = \left[(L - L_0)^2 + (a - a_0)^2 + (b - b_0)^2\right]^{0.5}$$

where L_0, a_0, and b_0 are the color parameters of the initial stage (or stage 1) and L, a, and b are the corresponding color parameters at other processing stages.

In the frying industry, color can be measured by a colorimeter or comparison charts. To duplicate the responses of the human eye, the colorimeter uses a set of three filters (red, green, and blue) with transmission curves similar to the standard X, Y, and Z curves. The light reflected from the object through each filter is recorded and the tristimulus values are obtained. All tristimulus colorimeters available depend on this principle with individual refinements in photocell response, sensitivity, stability, and reproducibility. Today, the meters also provide the color values in other color spaces. Tristimulus colorimeters usually are small and portable. They come with special attachments, a wide range of aperture openings, and are customized for specific applications. The colorimeters are available from Agtron Inc., Sparks, NV; BYK-Gardner, Silver Springs, MD; HF Scientific Inc., Ft. Myers, FL; Hunter Lab, Reston, VA; Minolta Corp., Ramsey, NJ; and X-Rite Inc., Grandville, MI.

Shyu et al. (2005) measured the surface color of vacuum-fried carrot chips with a colorimeter (Nippon Denshoku 90 color difference meter, Tokyo, Japan). They expressed color as Hunter L, a, and b values. Garayo and Moreira (2002) used a Hunter Lab Colorimeter Labscan XE (Hunter Associates Laboratory, Reston, VA) to measure color of vacuum-fried potato chips. Both Innawong (2001) and Ballard (2003) used a Minolta chromameter (Model CR-300, Minolta Camera, Ltd., Osaka, Japan) to measure the L*, a*, and b* values of chicken nugget crusts. Moyano et al. (2002) used this same instrument to determine French fry color. Baik and Mittal (2003) used it for fried tofu, and Ramirez and Cava (2005) used it for fried pork chop loins. Patterson et al. (2004) used a Minolta chromameter Model CR-200 for *akara* (fried cowpea paste). Dogan et al. (2005) used a Minolta color reader (CR-10) on fried chicken nuggets. Hue angle was used to characterize color.

8.4   JUICINESS

Juiciness is another important quality for breaded fried food products. In breaded fried foods, the combination of fat in the crust layers and moisture in the core affects the juiciness of the product when released inside the mouth. Juiciness in breaded fried food can be defined as the amount of juice released during consumption. This affects the texture, flavor, and overall acceptance of the product. The product should not be dry, over-exudative, or soggy. Juiciness can be measured by sensory methods or objective methods using mechanical forces, as described by Mallikarjunan and Mittal (1994). In the press method, a set of pre-weighed filter papers and 0.5 to 1.5 g of sample is placed between two sheets of aluminum foil. This set of aluminum sheets is placed between filter papers and a set of Plexi-glass plates. The sandwich is then compressed by applying a 20-kPa pressure for 1 min using a laboratory press or universal testing machine. After pressing, the sample along with the aluminum foil is discarded. The set of filter papers is reweighed to obtain the mass of pressed juice. The weight increase of the filter papers is then correlated to the expressed juice from the product.

Innawong (2001) measured juiciness of chicken nuggets using a press method described by Mallikarjunan and Mittal (1994), in which the core of the nugget was pressed with 20-kPa pressure for 1 minute. The juiciness was evaluated from the expressed juice collected by the filter paper. Higher amounts of expressed juice is correlated with a juicier product and Innawong (2001) concluded that using nitrogen as the pressurizing medium for frying chicken nuggets provided a juicier product compared to the conventional method of using steam released from the product as the pressurizing medium.

8.5   CRISPNESS

Crispness is one of most important textural and desirable characteristics of dry crisp foods and fried foods, and has been studied by many investigators. Various definitions and meanings of crispness, studies of instrumental measurement of crispness, and its importance are described in this section.

Crispness is a highly valued and universally liked textural characteristic that has many positive connotations. Its presence signifies freshness and high quality. It goes well with many other textural characteristics and is often used to create pleasing textural contrasts. Probing into consumer attitudes to texture and its specific characteristics, Szczesniak and Kahn (1971) stated that crispness is particularly good as an appetizer and as a stimulant to active eating. It is very important to the pleasure of substantial eating. It appears to hold a particular place in the basic psychology of appetite and hunger satiation. It is notable as a relaxing or satiable texture and appears to be a universally liked characteristic. Crispness is very prominent in texture combinations that mark excellent cooking and is synonymous with freshness and wholesomeness. In their 1984 study of textural combinations, Szczesniak and Kahn added that crispness appears to be the most versatile single texture parameter.

Over the years, many investigators have worked on the various definitions and meanings of crispness. However, the definition of crispness is not completely understood. Only the generalized concept has been established as close to the definition as possible. The importance and desirability of crispness have increased research efforts to define and measure this attribute (Vickers, 1988; Szczesniak, 1988; Lee, Schweitzer, Morgan, and Shepherd, 1990; Dacremont, 1995). Szczesniak (1988) tried to characterize crispness based on the consumer descriptions and found that crispness was associated with brittleness, crackling, snapping, crunchiness, and sound emission during eating.

8.5.1 BATTERS AND BREADING FOR FRIED FOOD AND THEIR SIGNIFICANCE TO CRISPNESS

Among battered and breaded food products, fried foods constitute a major portion. Fried chicken products exceeded $8.2 billion in sales in the United States in 1996. Coating seafood, poultry, red meat, and vegetable products with a batter and breading before cooking is a common practice of homemakers, food processors, and commercial fast food outlets. Batter is defined as a liquid mixture comprised of water, flour, starch, and seasonings into which food products are dipped prior to cooking. Breading is defined as a dry mixture of flour, starch, and seasonings, coarse in nature, and applied to moistened or battered food products prior to cooking. Coating is referred to as the batter and breading adhering to a food product after cooking (Suderman, 1983).

The batter uniformity and thickness, which is related to the batter viscosity, determine acceptability of the finished product. A more viscous batter will pick up more breading than will a less viscous one.

Batters and breadings serve many functions as food coatings, such as enhancing a food product's appearance (Elston, 1975), giving a crispy texture (Elston, 1975 and Zwiercan, 1974), and contributing to the pleasure of substantial eating (Vickers and Bourne, 1976b). Coating material is a key for producing a desirable crispness in breaded fried chicken nuggets, chicken strips, and seafood. Ideally, the coating should exhibit a structure that sufficiently resists the initial bite and then should disappear with a quick melt away in the mouth. A coating that does not readily break down during subsequent mastication will be rated chewy, heavy, undesirable, and perhaps even lacking in freshness (Loewe, 1992).

Hanson and Fletcher (1963) suggested that mixtures of thickening agents could be used for achieving desirable crispness. Cooking also affects coating crispness. Donahoo (1970) reported that crispness could be adjusted by time and temperature of cooking.

The optimum method for producing crisp coated foods is through deep-fat frying at temperatures ranging from 176 to 204°C. Rapid heat transfer quickly sets the coating structure, allowing little time for excess moisture infiltration. This cooking procedure is the method of choice in the food service industry for both interface/adhesion and puff/tempura coatings. Primarily used in the home, oven heating is the method for producing a moderately acceptable product in terms of crispness, color, and flavor. Although the heating rate is slower than that of deep-fat frying, the elevated chamber temperature of an oven causes some evaporative drying of the coating, resulting in the perception of crispness. It appears that microwave heating is not a suitable method for coated foods. Microwave oscillations cause molecular vibrations and resultant frictional heating within the food. The evaporative drying does not occur and the result is a soggy coating with minimal crispness. Use of microwave heating will require a unique technology for effective product development of coated foods (Kulp and Loewe, 1992).

One problem common to all battered and breaded products is adhesion. Poultry parts are known as the most difficult food substrate to batter or bread (Suderman, 1983). Batter and breading adhesion is affected by several factors such as poultry skin ultrastructure, freezing of parts, product surface temperature, predip composition, batter viscosity, breading and batter composition, and the cooking process. In typical deep-fat frying, the batter or breading coating quickly coagulates upon exposure to high frying-oil temperatures. As a result, the coating essentially takes the size and configuration of the product. However, as the food product continues to cook, the substrate shrinks to a size smaller than the coating matrix. Precooking of the product by steaming, simmering, or boiling has been recommended to improve its adhesive properties before the batter is applied (Kulp and Loewe, 1992).

8.5.2 CRISPNESS AND CRUNCHINESS

Several scientists have shown that crispness and crunchiness are very closely related sensations, but crunchiness is used more often in reference to moist foods (Szczesniak and Kahn, 1971; Moskowitz and Kapsalis, 1974; Vickers and Wasserman, 1980; Vickers, 1981). Moskowitz and Kapsalis (1974) derived regression equations relating descriptors to one another and found that crispness was most closely related to the quality of crunchiness and crunchiness was most closely related to crispness and hardness. Vickers and Wasserman (1980) used multidimensional scaling to arrange 15 food sound descriptors in two-dimensional space. The descriptors crisp and crunchy were very close to each other in this space, indicating that they have similar meanings when used to describe sounds. Vickers (1981) had panelists judge 16 foods and recorded the biting and chewing sounds of these foods for both crispness and crunchiness. She found large correlations between the two descriptors, whether the judgments were made based on biting and chewing the foods or by only listening to the sounds.

It has been suggested that the sensations of crispness and crunchiness may differ in the pitch of their respective sounds (Vickers, 1979, 1984). Foods that were more crisp rather than crunchy produced higher pitched sounds than foods that were more crunchy than crisp. However, pitch is a complex characteristic and is not dependent on a single physical quantity, the pitch of sound being determined by its frequency, intensity, and waveform.

Seymour and Hamann (1988) studied the relationships between descriptive sensory crispness and crunchiness and acoustic and mechanical measurements for low moisture foods. A trained texture profile panel developed sensory definitions of crispness and crunchiness. Sensory crispness was evaluated by placing samples between incisors and detecting a level of higher pitched noise. Sensory crunchiness was evaluated by placing samples between molar teeth and detecting a degree of low-pitched noise. High correlations were found between sensory crispness and crunchiness in all products. Seymour and Hamann (1988) indicated that crispness and crunchiness are closely related sensory interpretations of food texture, and can be quantified by a combination of mechanical and acoustic measurements. Mechanical force and work done to failure had strong inverse correlations with sensory crispness and crunchiness. Acoustic parameters also had high correlations with sensory parameters. Crisp products were mechanically weaker than crunchy products. More force was required to fracture a crunchy product. However, the acoustic parameters that highly correlated with sensory crunchiness tended to be at lower frequencies than was the case for crisp products.

8.5.3 CHARACTERIZATION AND DETERMINATION OF CRISPNESS

Although there are many approaches to the instrumental measurement of crispness in foods, the best measurements are still inconclusive. However, the properties related to crispness were able to disclose the complexity of crispness and its association to other similar sensory attributes, such as brittleness, hardness, crackliness, or crunchiness.

8.5.3.1 Structural and Geometrical Properties

Many researchers agree that crispness should result from the structural properties of a food (Barrett, Rosenberg, and Ross, 1994; Gao and Tan, 1996; Bouvier, Bonneville, and Goullieux, 1997; Mohammed, Jowitt, and Brennan, 1982; Stanley and Tung, 1976; Vickers and Bourne, 1976b). Matz (1962) and Coppock and Carnford (1960) indicated that crisp, dry foods such as biscuits break into many pieces when masticated and that their eating quality is affected by the size of air cells and thickness of the cell walls.

Crispness is related to the cellular structure of foods. Perhaps the most direct method of its objective measurement is likely to be the investigation of the product's structure and geometrical properties. Scanning electronic microscopy (SEM) is often used to reveal the internal structure of the product. Gao and Tan (1996) used this technique to measure the cell size and density. Barrett et al. (1994) investigated the structural properties characterized in terms of cell size distribution and bulk density of corn-based extrudates. They found that mechanical strength, defined by a compressive stress, and fracturability, quantified by fractal and Fourier analyses of stress–strain functions, increased with either decreasing mean cell size or increasing bulk density. The correlation of fracturability parameters or structural characteristics with sensory scores of crunchiness, crispness, and hardness indicated that cellular structure strongly influences the pattern of mechanical failure.

Recently, Gao and Tan (1996b) proposed that some important sensory attributes could be analyzed by an image processing technique. Some important sensory attributes could be predicted by processing the surface and cross-section images of the product.

8.5.3.2 Mechanical Properties

Perhaps the most prevalent objective measurement for crispness is a determination via mechanical properties. The mechanical properties are associated with the structural properties of materials derived by means of the resistance to a compression of blade/probe and to a tensile that pulls the structure of food material apart by a universal testing machine such as Instron or a texture analyzer.

In order to authenticate the sensory assumptions, various modifications of jigs and tools were created for objective investigations, such as shear compression blade, puncture probe, Kramer shear-compression test cell, and snap test cell. Nevertheless, there are no definite criteria for selection of an apparatus to measure the mechanical properties of foods. In addition, the tests are dependent upon the nature of products. Therefore, a variety of mechanical tests have been reported for different low-moisture foods.

Vickers and Bourne (1976a) used a snap test to measure parameters such as bend deformation to fracture, and stiffness (the slope of a force-deformation curve of Young's modulus). They found large correlations between these measurements and sensory crispness. Voisey and Stanley (1979) suggested that the number of peaks or breaks using a Warner-Bratzler test cell would be a good indicator of crispness in fried bacon.

Mohamed et al. (1982) used a constant force rate texture-testing instrument to study crispness of the biscuits. Good correlations were found between sensory crispness and the ratio of work to fracture to total work. Seymour (1985) used a Kramer shear cell in an Instron to crush samples of several dry crisp foods altered in crispness by humidification. He found large negative correlations between crispness and the following mechanical parameters: maximum force at failure and work done to failure.

Although mechanical tests are relatively quick and easy to perform, they have not produced a high enough degree of correlation with sensory crispness. In addition, many crisp foods cannot be tested by these tests because they are too small, have irregular sizes and shapes, or are part of a food that also consists of non-crisp parts.

8.5.3.3 Acoustical Properties

Perhaps the first impression of a crisp food is the sound burst during biting. Because the crushing of crisp or crunchy foods results in fracture and fragmentation, it appears that fracture and sound emission are associated. Attenburrow, Davies, Goodband, and Ingman (1992) reported that the sounds emitted during the crushing of a dry product are due to a sudden release of stored elastic energy. Al Chakra, Allaf, and Jemai (1996) have further elaborated the association between mechanical fracture and sound emission via the first and second laws of thermodynamic principles. In brief, the rupture of a brittle product, obtained as the applied stress reaches a critical value, induces an instantaneous liberation of the elastic energy from a binding of the inter-atomic bonds, in the form of acoustical energy.

Pioneering work on acoustical properties in foods was conducted by Drake (1963, 1965), who found that sounds from crisp foods differ from non-crisp foods in loudness. Later in the 1970s and 1980s, acoustic measurement in food research gained more attention in the characterization of sound-related textural attributes, e.g., crispness and crunchiness. Mechanical properties were primarily used to determine and interpret crispness until Vickers (1987), Seymour and Hamann (1988), Mohamed et al. (1982), and Vickers and Bourne (1976a) found that the combination of the sound and mechanical properties was better in predicting sensory crispness than the mechanical properties alone. For example, crispness was found to be poorly correlated with an instrumental fracture force (r = 0.018) in a low-moisture food (Mohamed et al., 1982). This indicated that mechanical properties alone might not entirely explain crispness. Sensory crispness has been predicted using both mechanical and acoustical variables by the multiple linear regression technique summarized in Table 8.1.

Studies have shown that the auditory sensations are important for evaluating crispness (Vickers and Bourne, 1976b; Christensen and Vickers, 1981; Mohamed et al., 1982; Edmister and Vickers, 1985; Lee, Deibel, Glembin, and Munday, 1988). Vickers and Bourne (1976a) studied the acoustical properties of tape-recorded biting sounds of wet and dry crisp foods. They found that crisp foods consist of an uneven and irregular series of noises and suggested that the repeated breaking or fracturing of food samples during biting and chewing produced these acoustical characteristics. Observing differences in amplitude-time plots between the samples, Vickers and Bourne (1976a) concluded that less-crisp samples produced less noise. Christensen and Vickers (1981) evaluated separately the loudness and crispness of 16 different products during chewing and biting. They found high positive correlations between crispness and loudness, indicating that biting and chewing sounds were important for evaluating crispness. Mohamed et al. (1982) studied the sound produced by five varieties of dry crisp foods stored at different relative humidity levels. The sounds were recorded as the foods were fractured by compressing in a constant loading rate texture-testing instrument. The sound energy correlated significantly with sensory crispness. Edmister and Vickers (1985) investigated the relationships between several instrumental acoustical parameters and sensory crispness. They found that the best acoustical predictor of auditory crispness was the logarithm of the number of sound bursts and the mean amplitude of the bursts. Vickers (1988) noted that an inverse relationship between crispness and the force-deformation variables (a negative sign preceding the force and work term) is more unusual and could mean that sensations of hardness and toughness are detracting from the sensation of crispness.

Another problem in acoustic analysis relates to the source of the sound generation. Dacremont (1995) asserted that sounds generated from a food fracturing through a mechanical apparatus are different from eating sounds and do not contain the relevant information for texture judgment. Nevertheless, eating sounds still comprise various frequency components, which can be either airborne or conducted via the bone. Regarding this matter, previously Lee et al. (1988) investigated acoustic behavior during 10 consecutive chews of potato chips and tortilla chips. They found that as chews increased, sound intensity tended to decrease. In addition, the higher frequency of chewing sound, which is audible, decreased as chews increased. The latter finding was supportive of the psycho-acoustical theory proposed by Vickers (1979) that crispness should be characterized by high-pitched sound. Therefore, she hypothesized that the assessment of crispness may be more dependent on the information obtained from initial mastication as opposed to later chews.

Peleg (1997) has raised cautions concerning the measured acoustic variables used in the past. Peleg specifically questioned the use of peak count because it does not account for the peak magnitude and shape. In addition, the count can also be affected by the selected resolution and sampling rate adopted by the researchers. Peleg (1997) recommended using the Fourier transform method to obtain information that is more reliable for determining crispness (or crunchiness) in foods.

The development of acoustic measurement occurred in the late 1990s. Al Chakra et al. (1996) suggested that structural properties of pasta could be characterized based on the sound emission during its rupture. The mechanical and acoustical parameters changed in the same pattern over the water activity ranges, and good correlation was obtained. This ascertained their proposed assumption that both properties would have an existing link with each other. Nevertheless, characterization appeared to be more complicated since Tesch, Normand, and Peleg (1996) found no relationships between mechanical and acoustical parameters. They were concerned that its complication may have been due to the effect of unequal frequency for the mechanical and acoustical measurements or that some parts of the crisp or crunchy information manifested in acoustical properties may not have been fully revealed in the mechanical properties, or vice versa. In response to the work of Al Chakra et al. (1996), Tesch et al. (1996) commented that changes in mechanical properties need not follow the same trend as that of acoustical properties, particularly if the changes occurred around the glass transition region. At this moment, acoustical measurements for crispness are undergoing study and are being used in the investigations into the effect of glass transition on tested materials. Hopefully, some textural changes in that range, particularly with regard to crispness, can be revealed and elucidate the links between the material itself and the mechanical and acoustical properties.

Tahnpoonsuk (1999) conducted sensory, instrumental, and acoustic measurements to evaluate crispness in breaded shrimp baked in an oven and held under a heat lamp. The panel, consisting of 12 trained panelists, rated the intensity of crispness on a 150-mm line scale with anchors at the ends. Crispness was defined as the ease of fracture in the mouth combined with loudness of the sound produced. Panelists perceived significant differences in crispness among the tested samples. The change in sensory crispness was significantly dependent on the holding time under a heat lamp and baking location in the oven. Although the objective analysis was significant, it did not produce satisfactory correlations with sensory crispness.

In addition to using sound in the hearing range by humans, Povey and Harden (1981) measured crispness of biscuits using the ultrasonic pulse echo technique. They found good correlation between the crispness from sensory measurement and the velocity of longitudinal sound. The ultrasonic velocity correlated with crispness better than either the ultrasonically derived Young's modulus, or the Instron universal testing machine derived modulus. Povey and Harden concluded that the ultrasonic technique offers promise as a method for the electromechanical measurement of crispness.

Antonova et al. (2003) attempted to characterize crispness in breaded chicken nuggets using an ultrasonic technique and found that the ultrasonic velocity had a higher correlation with sensory crispness. To create a range of crispness values, they used different finish cooking methods (frying, baking, and microwave cooking) and held the samples under a heat lamp set at 60°C for a period of 40 min. The samples were examined using a high power ultrasound system in a through transmission mode. The pulse width and separation time between pulses were adjusted for 250-kHz transducers. The velocity of the ultrasound was obtained to represent the crispness in the product.

Ultrasonic velocity represents the average speed of ultrasound through the sample from one side to the other. The time-of-flight (TOF), known as the traveling time of the ultrasonic pulse from one side of sample to the other, was derived from the plot of the time-domain waveform. The predetermined sample thickness and the determined TOF through the sample and TOF of the transducer were used to calculate the propagation velocity of the ultrasonic wave through the sample.

The ultrasonic velocity for each breaded fried chicken nugget was determined according to the following equation:

$$\upsilon_{sample} = \frac{l}{TOF - TOF_0} \tag{8.1}$$

where

υ_{sample} = Ultrasonic velocity for breaded fried chicken nugget (m/sec)
l = Path length of transmission (mm)
TOF = Time-of-flight with the sample (ms)
TOF_0 = Calibrated time-of-flight without the sample (ms)

A trained sensory panel was used to obtain sensory crispness and was correlated with ultrasound velocity. The panelists were trained using the standard rating scale (Table 8.1) to characterize crispness in breaded fried foods. The samples cooked in a deep-fat fryer had higher ultrasonic velocities (431.56 to 715.38 m/sec) than did the samples cooked in an oven (221.41 to 533.83 m/sec), while the samples cooked in a microwave oven had much lower velocities (90.22 to 306.92 m/sec). Higher ultrasonic velocities correlated with higher crispness ($R^2 = 0.83$) in the samples, suggesting that ultrasound traveled faster in dry and crisp samples compared to moist and soggy ones (Figure 8.1). Antonova et al. (2003) also found very high correlation between moisture content of the samples with sensory crispness and ultrasonic velocity.

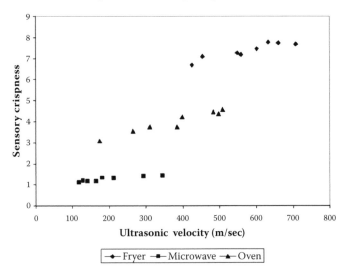

FIGURE 8.1 Correlating sensory crispness in the breaded fried food to ultrasonic velocity. *Source*: Antonova, I., Mallikarjunan, P. and Duncan, S.E. 2003. Sensory assessment of crispness in a breaded fried food held under heat lamp. *Foodserv. Res. International* 14(3): 189–200.

TABLE 8.1
Standard Rating Scale for Crispness during Panel Training

Product	Brand/Type/Manufacturer	Standard Value
Granola Bar	**Quaker Low Fat Chewy Chunk**	1.0
Club Cracker	**Keeblers Partner Club Cracker**	3.0
Graham Cracker	**Honey Maid**	3.0
Oat Cereal	**Cheerios**	4.0
Bran Flakes	**Kellogg's Bran Flakes Cereal**	5.0
Cheese Crackers Goldfish	**Cheddar Cheese Crackers**	6.0
Corn Flakes	**Kellogg's Corn Flakes Cereal**	7.0
Melba Toast	**Devonsheer Melba Toast**	9.0

Rating: 1 = soggy/not crisp; 9 = very crisp

8.5.3.4 Combination Measurements

Sensory crispness has been predicted using both acoustical and force-deformation measurements by the multiple linear regression technique summarized in Table 8.2. Seymour (1985) studied crispness in four different low moisture foods. In all cases except Crunch Twist, crispness was positively related to an acoustical parameter and inversely related to a force-deformation parameter. This inverse relationship of crispness to force-deformation parameters and positive relations to acoustical parameters was also shown by Vickers (1987, 1988). Vickers (1988) noted that the inverse relationship between crispness and the force-deformation parameters is more unusual and may mean that sensations of hardness and toughness are detracting from crispness.

Characterization of crispness has been more complicated since Tesch et al. (1996) found no relationships between mechanical and acoustical parameters. They were concerned that complications may have been due to the effect of unequal frequency for the mechanical and acoustical measurements or that some parts of the crisp or crunchy information manifested in acoustical properties may not have been fully revealed in the mechanical properties, or vice versa.

However, all of these tests were applied for low-moisture foods only. Studies related to crispness in high-moisture products have been very limited. In addition, the measurements may be very different from that developed for low-moisture foods. Recent efforts by Tahnpoonsuk and Hung (1998) to characterize crispness in breaded shrimp did not give a conclusive correlation. They attempted to record sound signals during compression testing, using a modified Warner-Bratzler shear blade (Figure 8.2).

TABLE 8.2
The Regression Equation for Crispness Prediction in Different Foods

Products	Regression Equation	R^2
Seymour (1985)		
Potato chips (Pringles)	Crispness = 13.6 − 0.19 Work^x + 0.03 MP4	0.96
Potato chips (O'Gradys)	Crispness = 6.1 − 0.08 Force + 0.21 SPL3	0.88
Potato chips (Rippled Pringles)	Crispness = 8.1 − 0.004 Work^y + 0.003 ILT	0.89
Crunch Twist	Crispness = 16.5 − 0.06 Force − 0.11 SPL1	0.95
Saltine crackers	Crispness = 9.5 − 0.12 Force + 0.017 MPL	0.91
Vickers (1987)		
Potato chips	Crispness = -15.6 + 5.35 NP + 133 MHP − 6.21 Peak	0.98
Vickers (1988)		
Breakfast cereals	Crispness = 538 + 539 (log MHP) − 222 Peak	0.74
Antonova et al. (2003)		
Chicken nuggets	Crispness = 3.35 (Peak) − 5.20	0.64
	Crispness = 0.013 (Velocity) − 0.67	0.83

Force = Maximum force at failure (N) by Kramer shear cell
Work^x = Work done to 1 cm deformation (mJ)
Work^y = Work done to failure (mJ)
MP4 = Mean sound pressure (N/m^2) in 2.6-3.3 kHz
SPL1 = Mean sound pressure level (dBA) in 0.5-1.2 kHz
SPL3 = Mean sound pressure level (dBA) in 1.9-2.6 kHz
ILT = Acoustic intensity (watts/m^2) in 0.5-3.3 kHz
MPL = Acoustic intensity (watts/m^2) in 0.5-1.9 kHz
MHP = Mean height peaks taken from oscilloscope display of bite sounds
NP = Number of sound occurrences during bite
Peak = Maximum force from a force-deformation curve
Velocity = Ultrasound velocity

8.5.3.5   Future Trends

As with any development of methods to characterize a complex textural attribute, like crispness, further work is needed to validate these methods by collaborative research on various types of instruments and products. Research is in progress by many researchers in describing crispness in relation to other measurable attributes like moisture content, and mechanical and acoustic properties. Further work is underway in relating various texture attributes to glass transition temperature. Novel numerical methods such as fractal analysis and Fourier transformation have been employed to describe the objective measurements, especially for the acoustic properties.

Research related to using ultrasonic techniques with air-coupled transducers will lead to a true non-destructive, non-contact, rapid crispness evaluation method. However, use of air-coupled transducers has been limited to low-power diagnostic measurement methods, while crispness measurement requires a high-power diagnostic system and might result in the use of a through-transmission system as developed

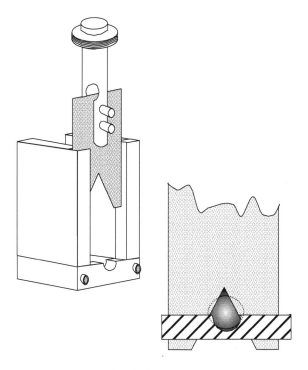

FIGURE 8.2 Warner-Bratzler shearing device.

by Antonova et al. (2003) and described in the earlier part of this chapter. Future research could also focus on the use of artificial neural network (ANN) techniques to process multiple attributes (such as moisture content, ultrasonic velocity, color, mechanical properties, and acoustic properties) from the sample to arrive at a comprehensive crispness evaluation that is well correlated with the sensory crispness.

8.5.4 PARAMETERS CONTROLLING CRISPNESS

8.5.4.1 Process and Structure

Barret, Cardello, Lesher, and Taub (1994) suggested that most cereal-based crispy products are brittle material characterized by a cellular, lamellar, or puffed structure. They also suggested a relationship describing crispness (through fractal dimension) with a combination of cell area (A) and bulk density (ρ) expressed as the following:

$$\text{Fractal Dimension} = 1.37 - 0.0112A + 1.93\rho$$

The relationships between process and crispness have been studied considering different processing conditions. Most of these studies are based on extrusion cooking, investigating the role of water content, screw speed, torque, pressure, and temperature. These parameters interact largely and their action on texture may be variable depending on the combination chosen. Simultaneous high shear and temperature give rise to crispness (Onwulata and Heymann, 1994).

The crispness development during the process has been investigated for frying. Kawas and Moreira (2001) reported that the crispness of chips increased as porosity increased and moisture decreased.

8.5.4.2   Ingredients and Hydration

Most low-moisture baked or deep-fried products have a crispy texture. If the moisture content of these products increases, due to water absorption from the atmosphere or by mass transport from neighboring component phases, a loss of crispness is observed (Nicholls et al., 1995). The effect of hydration on crispness can be described by a Fermi equation expressed as:

$$Y = Y0/(1 + \exp[(aw - awc)/b])$$

where Y is crispness, Y0 is crispness in the dry state, awc is the critical water activity corresponding to $Y = Y0/2$, and b is a constant that express the transition range.

Roudaut et al. (1998) reported that crispness of cereal-based products could be affected by hydration at temperatures below their Tg. Nikolaidis and Labuza (1996) attributed hydration-induced texture changes to the glass transition of gluten. Due to the complexity of the products, products may contain multiple phases with different Tg's. The textural changes could thus be caused by the glass transition of a minor phase, which may not be visible on DSC thermograms. However, such a point may not be valid (Slade and Levine, 1993).

Since ingredients affect the structural organization of products, they are likely to control their mechanical properties and most expectedly their crispness (Onwulata et al., 2001). The effect of sucrose on crispness has been investigated in various ways. Nussinovitch et al. (2000) observed that, for freeze-dried agar prepared with or without infused sucrose, sucrose increased brittleness (and thus expectedly crispness), although the interpretation was ambiguous for how sucrose affected the density of the material. Sucrose exhibits a crispness protecting action against hydration, shifting for sucrose-rich extruded starch the critical hydration toward values higher than those for pure extruded starch (Roudaut et al., 2001). Ferriola and Stone (1998) considered the influence of secondary sweeteners (coating) on the "bowl life" of crispness, with significant "wet" crispness stabilization by honey.

8.5.4.3   Others

Pre-treatment. Krokida (2001) reported air-drying pre-treatment increased the crispness of potato strips, while osmotic pre-treatment did not affect the crispness significantly, with the exception of sugar pre-treated samples.

Frying-media. Frying oil contributes an important role to the food crispness as well. Krokida et al. (2001) pointed out that the crispness of potato strips is higher when fried in hydrogenated oil, while the use of refined oil (cottonseed oil) decreases the crispness.

8.6 OTHER TEXTURAL ATTRIBUTES

One of the most important quality attributes of fried food, having a major influence on its popularity, is texture. Fried food texture depends mainly on the ingredients in the batter and breading. Textural properties are related to the deformation, disintegration, and flow of food under force. These properties can be measured objectively and can be correlated to sensory textural attributes.

Szczesniak (1963) gave a system of classification of textural characteristics on fundamental rheological principles. Textural characteristics can be classified into five primary parameters of hardness, cohesiveness, viscosity, elasticity, and adhesiveness, and then into three secondary parameters of brittleness, chewiness, and gumminess. These properties are referred to the manner in which the food behaved in the mouth.

Tenderness has been measured as shear, bite, penetration, tensile, and compressive forces. Some objective methods of evaluating the tenderness in fried foods include shear methods, using Warner-Bratzler shear device or Kramer shear press, and compression methods (puncture test, texture profile analysis).

In measuring texture, force must be recorded accurately and since rapid force fluctuation and slope changes occur, a high-frequency response capability for the data acquisition is required. In addition, accurate records of the probe position are needed to interpret the data properly. A reproducible deformation rate is critical. In order to achieve a faster data processing and reproducible deformation, use of a universal testing machine for conducting the experiments is necessary. The universal testing machines are available in the market from Instron Corp., Canton, MA; TA-Instruments Inc., New Castle, DE; Testing Machines Inc., Amityville, NY; and Texture Technologies Corp., Scarsdale, NY.

8.6.1 SHEAR TEST

One of the widely used texture measurements is the Warner-Bratzler shear test. In this test, the sample is sheared using a Warner-Bratzler shear blade. The shear blade has an opening in the shape of an inverted triangle. The sample is placed inside the hole and the shear blade is moved in an upward direction during testing (Figure 8.1). Detailed discussion on the deformation pattern during testing and interpretations of the testing is widely available in the literature (Voicey, 1976).

The following parameters are obtained during a Warner-Bratzler shear test:

1. Maximum or initial yield force representing a tensile rupture signifying cohesiveness.
2. Cross-sectional area of the sample at rupture, which indicates compression required to initiate rupture and can be correlated with firmness.
3. Slope of the force–time curve that provides an index of firmness.
4. Force per unit length of the blade edge cutting the sample at the peak force.

Another shear test that is widely used is the Kramer shear test. In the Kramer shear test, the sample is placed in a Kramer shear cell (Figure 8.3). The Kramer shear cell contains

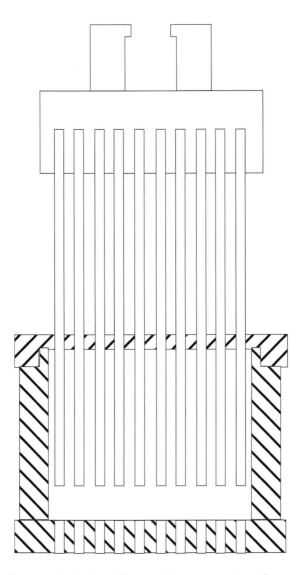

FIGURE 8.3   Cross-sectional view of Kramer shear compression cell.

a sample holder and a blade assembly. The Kramer shear blade assembly contains 10 blades, each 0.3175 cm thick. The sample holder is a metal box (6.7 × 6.7 × 6.3 cm) having 10 slots in its bottom. The sample is filled in the sample holder to the desired level (normally to 30% by volume). The sample holder has a matching slotted cover with 10 slots. The cover is placed on the top of the sample holder. The whole assembly is placed in the base of the universal testing machine. The blade assembly is lowered so that the bottom of the set of blades is just touching the slotted cover. After setting the universal testing machine to the desired test speed, the blades are lowered until the product is extruded through the bottom slots. A typical test speed is approximately 225 mm/min (22.5 cm/min).

Similarly, the following parameters are obtained during a Kramer shear test:

1. The initial yield force and yield distance
2. Peak force and peak force distance
3. Slope and work (the area under the force–time curve) required to shear the sample

Yang and Chen (1980) studied the effect of holding fried chicken in an oven on the textural quality using Warner-Bratzler shear test. The texture measurements were obtained on cooked samples. Biswas, Keshri, and Chidanandaiah (2005) used a Warner-Bratzlershear apparatus to determine the effects of different cooking methods on the quality of fried pork patties. Modi, Mahendrakar, Sachindra, and Rao (2004) compared the textural quality of nuggets made from fresh and smoked chicken meat using the Warner-Bratzler shear method.

8.6.2 Puncture Test

The puncture test is popular due to its simplicity because only the force required to push a puncture probe into or through the sample is to be determined. Standard penetration techniques with various shapes of probes have been used. Depending on the size and shape of the probe, the puncture test may only give data related to the sample compressibility and the force necessary to break down the sample. In addition to using a puncture probe, the sample can be placed on a die so the probe can penetrate through the sample completely.

The following three parameters are obtained during a puncture test:

1. Maximum or initial yield force that can be correlated with firmness
2. Sample area at rupture, which indicates energy required to initiate rupture
3. Slope of the force–deformation curve that provides modulus of elasticity

It is very common to see the use of a puncture test to describe crispness in fried foods. Many authors (Pinthus, Weinberg, and Saguy, 1995; Tan, Hung, and McWatters, 1995; Rovedo, Zorrilla, and Singh, 1999; Agblor and Scanlon, 2000; Rayner et al., 2000; Pedreschi, Aguilera, and Pyle, 2001; García et al., 2002; and Pedreschi and Moyano, 2005) have used the puncture test as a means to describe textural attributes in fried foods.

8.6.3 Textural Profile Analysis

Cyclic loading and unloading to a set deformation during uniaxial compression has been widely used by many investigators to determine several textural parameters. Using texture profile analysis, seven different texture properties can be estimated: (1) hardness, (2) brittleness, (3) cohesiveness, (4) elasticity, (5) adhesiveness, (6) chewiness, and (7) gumminess. Figure 8.4 illustrates a typical texture profile curve obtained using a universal testing machine.

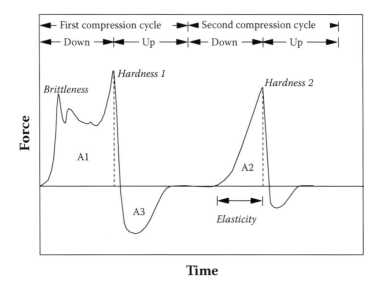

FIGURE 8.4 Typical force–time curve during a texture profile analysis.

Hardness is measured from the profile as the height of the first peak. Cohesiveness is measured as the ratio of the area under the second peak and the area under the first peak. Springiness (or elasticity) is defined as the height that the food recovers during the time that elapses between the end of the first bite and the start of the second bite. Adhesiveness is measured as the area, A3, of the negative peak beneath the base line of the profile and represents work necessary to pull the probe from the sample. Brittleness is characterized by the multi-peak shape of the first bite and is measured as the height of the first significant break in the peak. Gumminess is expressed as the product of hardness and cohesiveness. Chewiness is expressed as the product of hardness, cohesiveness, and elasticity.

Dogan et al. (2005) determined texture of chicken nuggets in terms of fracturability or brittleness, which is defined as the force at the first significant break in the first positive bite area of the texture profile analysis curve. A conical probe was attached to a texture analyzer (Lloyd Instruments, TA Plus, Hants, UK). Shyu et al. (2005) used a TA-XT2 texture analyzer (Stable Micro Systems Co. Ltd., Godalming, Surrey, UK) to determine the breaking force of vacuum-fried carrot chips. Pedreschi and Moyano (2005) did the same with potato chips. Kita and Lisinska (2005) determined the texture of French fries using an Instron 5544 instrument with a rectangular attachment. The velocity of the head was 250 mm/min with a 100-kg load cell. Maximum shear force was determined. Patterson et al. (2004) used an Instron universal testing machine (model 1122, Instron Corp., Canton, MA) with a Kramer cell to characterize the texture of akara. They determined that peak force required a cut through the akara ball using a 500-kg load cell with a crosshead speed of 50 mm/min. Garayo and Moreira (2002) also used a TA-XT2 texture analyzer for a rupture test on vacuum-fried potato chips. Innawong (2001) used a Kramer shear unit with a Sintech/MTS universal testing machine (model 5G, MTS, NC) to characterize the texture of chicken nuggets. Peak load, total energy to peak load, and energy to failure point were calculated.

8.7   REFERENCES

AACC. 1986. Approved methods of the American Association of Cereal Chemists. Minneapolis, MN: AACC.

Agblor, A. and M. Scanlon, 2000. Processing conditions influencing the physical properties of French fried potatoes. *Potato Res.* 43(2):163–177.

Al Chakra, W., Allaf, K., and Jemai, A.B. 1996. Characterization of brittle food products: Application of the acoustical emission method. *J. Texture Stud.* 27(3): 327–348.

Antonova, I., Mallikarjunan, P., and Chinnan, M.S. 2002. Protein-based edible coatings for fried foods. In *Protein-based films and coatings.* Gennadios, A. (Ed.). Boca Raton, FL: CRC Press.

Antonova, I., Mallikarjunan, P. and Duncan, S.E. 2003. Sensory assessment of crispness in a breaded fried food held under heat lamp. *Foodserv. Res. International* 14(3): 189–200.

AOAC. 1980. *Official methods of analysis,* 13th ed. Washington, DC.

AOAC. 1984. *Official methods of analysis,* 14th ed. Washington, DC.

AOAC. 1995. *Official methods of analysis of AOAC international.* 16th ed., vol. 2, Arlington, VA.

AOAC. 2000. *Official methods of analysis of AOAC international,* 17th ed. Washington, DC.

Attenburrow, G.E., Davies, A.P., Goodband, R.M., and Ingman, S.J. 1992. The fracture behaviour of starch and gluten in the glassy state. *J. Cereal Sci.* 16: 1–12.

Baik, O.D. and Mittal.G.S. 2003. Kinetics of tofu color changes during deep-fat frying. *Lebensm.-Wiss. U.-Technol.* 36: 43–48.

Ballard, T. 2003. Application of edible coatings in maintaining crispness of breaded fried foods. Masters Thesis. Virginia Polytechnic Institute and State University.

Barret, A.H., Cardello, A.V., Lesher, L.L., and Taub, I.A. 1994. Cellularity, mechanical failure, and textural perception of corn meal extrudates. *J. Texture Stud.* 25: 77–95.

Barret, A.H., Rosenberg, S., and Ross, E.W. 1994. Fracture intensity distributions during compression of puffed corn meal extrudates: Method for quantifying fracturability. *J. Food Sci.* 59: 617–620.

Biswas, A.K., Keshri, R.C., and Chidanandaiah. 2005. Effect of different cooking methods on quality of enrobed pork patties. *J. Food Sci. Technol.* 42(2):173–175.

Bligh, E.G. and Dyer, W. 1959. A rapid method of total lipid extraction and purification. *Can. J. Biochem. Physio.* 37: 911–917.

Bouvier, J.M., Bonneville, R., and Goullieux, A. 1997. Instrumental methods for the measurement of extrudate crispness. *Agro-Food-Industry Hi-Tech.* 8(1): 16–19.

Christensen, C.M. and Vickers, Z.M. 1981. Relationships of chewing sounds to judgments of food crispness. *J. Food Sci.* 45: 574–578.

Coppock, J.B.M. and Carnford, S.J. 1960. Texture in foods, *SCI Monograph,* No. 7, p. 64.

Dacremont, C. 1995. Spectral composition of eating sounds generated by crispy, crunchy and crackly foods. *J. Texture Stud.,* 26: 27–43.

Dogan, S.F., Sahin, S., and Sumnu, G. 2005. Effects of soy and rice flour addition on batter rheology and quality of deep-fat fried chicken nuggets. *J. Food Eng.* 71(1):127–132.

Donahoo, P. 1970. Choosing the right batter and breading. 7th Proceedings of the Annual Poultry and Egg Further Processing Conference. Ohio State University, Columbus, Ohio.

Drake, B.K. 1963. Food crushing sounds — an introductory study. *J. Food Sci.* 28:233–241.

Drake, B.K. 1965. Food crushing sounds: comparisons of objective and subjective data. *J. Food Sci.* 30:556–559.

Edmister, J.A. and Vickers, Z.M. 1985. Instrumental acoustical measures of crispness in foods. *J. Texture Stud.* 16:153–167.

Elston, E. 1975. Why fish fingers top the market. *Fishing News Intl.* 14:30.

Ferriola, D. and Stone, M. 1998. Sweetener effects on flaked millet breakfast cereals. *J. Food Sci.* 63: 726–729.

Gao, X. and Tan, J. 1996a. Analysis of expanded-food texture by image processing. Part I: geometrical properties. *J. Food Process. Eng.* 19:425–444.

Gao, X. and Tan, J. 1996b. Analysis of expanded-food texture by image processing. Part II: mechanical properties. *J. Food Process Eng.* 19:445–456.

Garayo, J. and Moreira, R. 2002. Vacuum frying of potato chips. *J. Food Eng.* 55: 181–191.

García, M.A., Ferrero, C., Bértola, N., Martino, M., and Zaritzky, N. 2002. Edible coatings from cellulose derivatives to reduce oil uptake in fried products. *Innovative Food Sci. Emerging Technol.* 3(4): 391–397.

Hanson, H.L. and Fletcher, L.R. 1963. Adhesion of coatings on frozen fried chicken. *Food Technol.* 17:793.

Holownia, K.I., Chinnan, M.S., Erickson, M.C., and Mallikarjunan, P. 2000. Quality evaluation of edible film-coated chicken strips and frying oil. *J. Food Sci.* 65(6): 1087–1090.

Innawong, B. 2001. Improving fried product and frying oil quality using nitrogen gas in a pressure frying system. Ph.D. Dissertation. Virginia Polytechnic Institute and State University.

Kawas, M.-L., and Moreira, R.G. 2001. Characterization of product quality attributes of tortilla chips during the frying process. *J. Food Eng.* 47: 97–107.

Kita, A. and Lisinska, G. 2005. The influence of oil type and frying temperatures on the texture and oil content of French fries. *J. Sci. Food Agric.* 85: 2600–2604.

Krokida, M.K., Oreopoulou, V., Maroulis, Z.B., and Marinos-Kouris, D. 2001. Effect of pretreatment on viscoelastic behavior of potato strips. *J. Food Eng.* 50(1): 11–17.

Kulp, K. and Loewe, R. 1992. *Batters and breadings in food processing.* American Association of Cereal Chemists, St. Paul, MN.

Lee III, W.E., Deibel, A.E., Glembin, C.T., and Munday, E.G. 1988. Analysis of food crushing sounds during mastication frequency — time studies. *J. Texture Stud.* 19:27–38.

Lee III, W.E., Schweitzer, M.A., Morgan, G.M., and Shepherd, D.C. 1990. Analysis of food crushing sounds during mastication: total sound level studies. *J. Texture Stud.* 19:27–38.

Loewe, R. 1992. Ingredient selection for batter systems. In *Batters and Breadings in Food Processing*, (Kulp, K. and Loewe, R., eds.), American Association of Cereal Chemists, St. Paul, MN.

Mallikarjunan, P. and Mittal, G.S. 1994. Meat quality kinetics during beef carcass chilling. *J. Food Sci.* 59: 291–294, 302.

Maskat, M.Y. and Kerr, W.L. 2004. Effect of breading particle size on coating adhesion in breaded, fried chicken breasts. *J. Food Qual.* 27: 103–113.

Maskat, M.Y., Yip, H.H., and Mahali, H.M. 2005. The performance of a methyl cellulose-treated coating during the frying of a poultry product. *Intl. J. Food Sci. Technol.* 40: 811–816.

Matz, S.A. 1962. *Food texture.* AVI Publishing Co. Inc., Westport, CT.

Modi, V.K., Mahendrakar, N.S., Sachindra, M.N., and Rao, D.N. 2004. Quality of nuggets prepared from fresh and smoked spent layer chicken meat. *J. Muscle Foods.* 15(3):195–204.

Mohammed, A.A.A., Jowitt, R., and Brennan, J.G. 1982. Instrumental and sensory evaluations of crispness: 1 — in friable foods. *J. Food Eng.* 1: 55–75.

Moskowitz, H.R. and Kapsalis, J.G. 1974. Psychophysical relations in texture. Presented at the Symposium on Advances in Food Texture, Guelph, Ontario.

Moyano, P.C., Rioseco, V.K., and Gonzalez, P.A. 2002. Kinetics of crust color changes during deep-fat frying of impregnated French fries. *J. Food Engr.* 54: 249–255.

Nicholls, R.J., Appelqvist, I.A.M., Davies, A.P., Ingman, S.J. and Lillford, P.J., 1995. Glass transitions and fracture behaviour of gluten and starches within the glassy state. *J. Cereal Sci.* 21: 25–36.

Nikolaidis, A. and Labuza, T.P. 1996. Glass transition state diagram of a baked cracker and its relationships to gluten. *J. Food Sci.* 61: 803–806.

Nussinovitch, A. Corradini, M.G., Normand, M.D. and Peleg, M., 2000. Effect of sucrose on the mechanical and acoustic properties of freeze dried agar, _ carrageenan and gellan gels. *J. Texture Stud.* 31: 205–223.

Onwulata, C. and Heymann, H. 1994. Sensory properties of extruded corn meal related to the spatial distribution of process conditions. *J. Sensory Studies* 9, 101–112.

Onwulata, C.I., Smith, P.W., Konstance, R.P., and Holsinger, V.H. 2001. Incorporation of whey products in extruded corn, potato or rice snacks, *Food Res. International*, 34(8): 679–687.

Patterson, S.P., Phillips, R.D., McWatters, K.H., Hung, Y-C., and Chinnan, M.S. 2004. Fat reduction affects quality of akara (fried cowpea paste). *Intl. J. Food Sci. Technol.* 39: 681–689.

Pedreschi, F. and Moyano, P. 2005. Oil uptake and texture development in fried potato slices. *J. Food Eng.* 70(4): 557–563.

Pedreschi, F., Aguilera, J.M., and Pyle, L. 2001. Textural characterization and kinetics of potato strips during frying. *J. Food Sci.* 66(2): 314–318.

Peleg, M. 1997. Effect of absorbed moisture on the mechanical properties of cereal foods, instant coffee, legumes and nuts. In the Proceedings of 5th Conference of Food Engineering a topical conference at the Annual Meeting of the American Institute of Chemical Engineers (AICHE). Paper no. 68d.

Pinthus, E.J., Weinberg, P., and Saguy, I.S. 1995. Deep-fat fried potato product oil uptake as affected by crust physical properties. *J. Food Sci.* 60(4): 770–772.

Povey, M.J.W. and Harden, C.A. 1981. An application of the ultrasonic pulse echo technique to the measurement of crispness of biscuits. *J. Food Technol.* 16: 167–175.

Ramirez, M.R. and Cava, R. 2005. Changes in color, lipid oxidation and fatty acid composition of pork loin chops as affected by the type of culinary frying fat. *Lebensm.-Wiss. U.-Technol.* 38(7): 726–734.

Rayner, M., Ciolfi, V., Maves, B., Stedman, P., and Mittal, G.S. 2000. Development and application of soy-protein films to reduce fat intake in deep-fried foods. *J. Sci. Food Agric.* 80(6): 777–782.

Roudaut, G., Dacremont, C. and LeMeste, M., 1998. Influence of water on the crispness of cereal based foods acoustic, mechanical, and sensory studies. *J. Texture Stud.* 29: 199–213.

Roudaut, G., Dacremont, C., Valles Pamies, B., Colas, B., and LeMeste, M. 2002. Crispness: a critical review on sensory and material science approaches, *Trends Food Sci. & Tech.*, 13(6–7): 217–227.

Rovedo, C.O., Zorrilla, S.E., and Singh, R.P., 1999. Moisture migration in a potato starch patty during post-frying period. *J. Food Proc. Preserv.* 23(5): 407–420.

Seymour, S.K. 1985. Studies on the relationships between the mechanical, acoustical and sensory properties in low moisture food products. Ph.D. Thesis, North Carolina State University, Raleigh, NC.

Seymour, S.K. and Hamann, D.D. 1988. Crispness and crunchiness of selected low moisture foods. *J. Texture Stud.* 19: 79–95.

Shyu, S-L., Hau, L-B., and Hwang, L.S. 2005. Effects of processing conditions on the quality of vacuum-fried carrot chips. *J. Sci. Food Agric.* 85: 1903–1908.

Slade, L. and Levine, H. 1993. The glassy state phenomenon in food molecules. In *The glassy state in foods*, Blanshard, J.M.V. and Lillford, P.J. (Editors), Nottingham University Press, Nottingham, pp. 35–102.

Stanley, D.W. and Tung, M.A. 1976. Microstructure of food and its relationship to texture. In *Rheology and texture in food quality.* de Man, J.M., Voisey, P.W., Rasper, V.F., and Stanley, D.W., (Eds.). AVI Publishing Co., Inc., Westport, CT.

Suderman, D.R. 1983. Use of batters and breadings on food products: A review. In: *Batter and breading technology.* Suderman, D.R. and Cunningham, F.E. (Eds.). Westport, CT: AVI Publishing Company, Inc. pp. 1–13.

Suderman, D.R. and Cunningham, F.E 1983. *Batter and breading technology*, Westport, CT: AVI Publishing.

Szczesniak, A.S. 1963. Classification of textural characteristics. *J. Food Sci.* 38: 385–389.

Szczesniak, A.S. 1988. The meaning of textural characteristics — crispness. *J. Texture Stud.* 19: 51–59.

Szczesniak, A.S. and Kahn, E.L. 1971. Consumer awareness and attitudes to food texture. *J. Texture Stud.* 1: 280–295.

Szczesniak, A.S. and Kahn, E.L. 1984. Texture contrasts and combinations. A valued consumer attribute! *J. Texture Stud.* 15: 285–301.

Tahnpoonsuk, P. 1999. Determination of crispness in breaded shrimp. M.S. Thesis, The University of Georgia, Athens, GA.

Tahnpoonsuk, P. and Hung, Y.C. 1998. Effect of moisture on the crispness of a heterogeneous multiplayer food. IFT Annual Meeting Technical Program Abstracts.

Tan, P., Hung, Y-C., and Mcwatters, K.H. 1995. Akara (fried cowpea paste) quality as affected by frying/reheating conditions. *J. Food Sci.* 60(6): 1301–1306.

Tesch, R., Normand, M.D., and Peleg, M. 1996. Comparison of the acoustic and mechanical signatures of two cellular crunchy cereal foods at various water activity levels. *J. Sci. Food Agric.* 67: 453–459.

Vickers, Z.M. 1979. Crispness and crunchiness of foods. In *Food texture and rheology.* Sherman, P. (Ed.). Academic Press, London.

Vickers, Z.M. 1981. Relationships of chewing sounds to judgments of crispness, crunchiness and hardness. *J. Food Sci.* 47: 121–124.

Vickers, Z.M. 1984. Crispness and crunchiness — a difference in pitch? *J. Text. Stud.* 15: 157–163.

Vickers, Z.M. 1987. Crispness and crunchiness. In *Food texture: instrumental and sensory measurement.* Moskowitz, H.R. (Eds.). Marcel Dekker, Inc., New York.

Vickers, Z.M. 1988. Instrumental measures of crispness and their correlation with sensory assessment. *J. Texture Stud.* 19: 1–14.

Vickers, Z.M. and Bourne, M.C. 1976a. A psychoacoustical theory of crispness. *J. Food Sci.* 41: 1158–1164.

Vickers, Z.M. and Bourne, M.C. 1976b. Crispness in foods — a review. *J. Food Sci.* 41: 1153–1157.

Vickers, Z.M. and Wasserman, S.S. 1980. Sensory qualities of food sounds based on individual perceptions. *J. Texture Stud.* 11: 319–332.

Voisey, P.W. and Stanley, D.W. 1979. Interpretation of instrumental results in measuring bacon crispness and brittleness. *J. Can. Inst. Food Sci. Technol.* 12(1): 7–15.

Yang, C.S. and Chen, T.C. 1980. Effect of oven holding on qualities of fried chicken parts. *J. Food Sci.* 45(3): 635–637.

Zwiercan, G.A. 1974. Case of the weeping pies (and others). *Food Eng.* 46: 79, 81.

9 Recent Technologies to Enhance the Quality of Fried Foods

With the increase in obesity and the association of obesity and other health concerns with fried food consumption, the fried food industry is trying to find technologies to limit the fat uptake in fried foods. In the 1990s, the industry attempted to develop low-fat and low-calorie products but found them to be less acceptable due to a compromise in taste by fat and sugar substitutes. The food industry is still trying to reduce the fat uptake in fried foods as a means to achieve a quality product comparable to the traditional fried products but which can be perceived as healthy for consumption.

We now live in a society with large choices of quality food where the consumers' appreciation has become one main criterion in their food choice along with nutrition and safety. The contribution of texture to the consumers' appreciation of a food product has been studied for a long time. Among that research, the importance of crispness was highlighted, especially in breaded fried foods. For example, word association tests in which consumers were asked to generate attributes related to a list of specific foods showed that the term "crisp" was mentioned more often than any other attribute (Rohm, 1990). The perception of the consumer related to fried foods has not been any different from the 1970s to the present. Iles and Elson (1972) found that food products were ranked in the same order for crispness and consumers' preference. Katz and Labuza (1981) also showed the importance of crispness in food acceptability. How to keep the crispness of food products, especially in fried products, has become a major problem that the food producer must face. To achieve this objective requires the knowledge of intrinsic parameters (physical, chemical, etc.) responsible for crispness. But crispness, like any other textural attribute, depends not only on rheological/mechanical characteristics exhibited by a product, but also on the consumers who identify the sensations perceived upon eating as relevant to crispness. Thus, to understand both the mechanisms underlying the perception of crispness and the meaning consumers give to the term "crispy" is relevant to developing novel fried foods (Roudaut et al., 2002). This chapter discusses the recent technologies in these fields in order to improve the quality of fried products, especially in the reduction of oil uptake and oil decomposition and increase in the crispness of products.

9.1 REDUCTION OF OIL UPTAKE OF FRIED PRODUCTS

Although the excessive consumption of fried products will lead to many diseases, such foods remain popular because of their good taste. In consequence, there has been interest in determining how to minimize oil absorption during production but maintain the organoleptic characteristics that are favorable to the consumer, such as crispness, color, and flavor (Saguy and Pinthus, 1995).

In order to control the absorption of oil, we must first understand the mechanism of oil uptake during frying. During deep-fat frying, water in the crust evaporates and leaves the food, so the water in the core migrates to the crust. Therefore, the crust has to remain permeable. The fact that the vapor leaves the food and the fat enters later is the reason why fat uptake is largely determined by the moisture content of the food (Mehta and Swinburn, 2001; Southern et al., 2000). Because oil can penetrate the place from where water has evaporated, oil penetration only happens where the temperature is high enough, i.e., in the crust layer. Research shows that oil hardly penetrates in the cooked core and that the microstructure of the crust is the main determining factor in oil uptake (Pinthus, Weinberg, and Saguy, 1995). It has been shown, using differential scanning calorimetric (DSC) analysis (Aguilera and Gloria, 1997), that in fried potatoes the crust contains six times the oil content of that of the core. Ufheil and Escher (1996) studied the dynamics of oil uptake during deep-fat frying of potato slices and they determined that most of the oil is absorbed when the slices are removed from the fryer. Moreira, Sun, and Chen (1997) found that the largest amount of oil penetrates the structure of tortilla chips during the cooling period and not during deep frying. They determined that only 20% of the total oil content was absorbed during deep-frying and approximately 80% of the oil was kept on the surface of the product. In addition, they found that almost 64% of the total oil content was absorbed during the cooling period. Accordingly, Moreira and Barrufet (1998) explained the mechanism of oil absorption during cooling in terms of capillary forces in products like tortilla chips. However, the oil uptake mechanism in breaded fried foods (having a crispy crust and a moist core) is much different from thin products like potato chips or tortilla chips. Pinthus and Saguy (1994) researched the relationship between the initial interfacial tension of a restructured potato product and various frying media and the medium uptake during deep-fat frying. They found that total oil uptake was higher for lower initial interfacial tension, reflecting the importance of wetting phenomena. There are many factors influencing oil absorption such as the structure of the product and pretreatment of the product. In the following text, we discuss the reduction of oil absorption in five parts.

9.1.1 PRE-FRY TECHNIQUES

Pre-treatment of frying food (osmotic pre-treatment, blanching, and others) can lower oil absorption during deep-fat frying (Koelsch, 1994; Krochta and de Mulder-Johnston, 1997; Williams and Mittal, 1999a). Rimac-Brncic, Lelas, Rade, and Simundic (2004) reported that oil absorption was decreased 27 to 28% in fried potato strips blanched in calcium chloride solution. Debnath, Bhat, and Rastogi (2003) found that pre-fry drying had a significant effect on the deep-fat frying of chickpea flour-based ribbon snacks (Figure 9.1).

As the pre-fry drying time was increased (0 to 90 min) before frying at 175°C, it resulted in a decrease in kinetic coefficient for moisture transfer (0.056 to 0.039 per second) as well as in kinetic coefficient for oil transfer (0.063 to 0.035 per second).

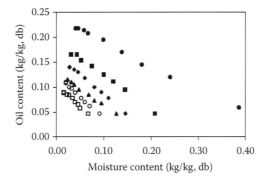

FIGURE 9.1   Variation of oil content against moisture content in frying of pre-fry dried ribbon snacks. Drying time ● = 0 min, ■ = 20 min, ◆ = 40 min, ▲ = 60 min, ○ = 90 min, and □ = 120 min. (Adapted from Debnath, S., Bhat, K.K., and Rastogi, N.K. 2003. Effect of pre-drying on kinetics of moisture loss and oil uptake during deep fat frying of chickpea flour-based food. *Lebensm.-Wiss.U.-Technol.* 36: 91–98.)

Based on the kinetics and sensory data, the optimum pre-fry drying times were found to be 40 min and 45 s, which resulted in 54% reduction in the oil content of the fried product. They thought the possible reason for the reduction in the oil content due to pre-fry drying could be the compactness of the material matrix (reduced porosity) or increase in the solid content.

9.1.2   POST-FRYING TECHNIQUES

Since most of the oil is taken up after removal of the product from the frying oil, the post-frying conditions are very important for fat uptake. The habits of the consumer during removal of the food from the fat can therefore play an important role. For instance, proper shaking and draining of the food are important techniques for reducing fat content of the food (Mellema, 2003). Mehta and Swinburn (2001) gave a good overview of the factors affecting fat uptake that can be controlled by the consumers. The main factors that they have associated with lower-fat content in potato chips included sample properties, pre-treatment, frying medium and conditions, and post-frying operations. In addition to sample thickness, surface characteristics, pre-treatment such as cryogenic freezing, sample moisture content, type of frying medium, sample-to-frying medium volume ratio, frying temperature, and post-frying procedures, additional measures that significantly reduce oil uptake in finished products include vigorous shaking of the fry-basket, hanging the fry-basket over the fryer to drain after frying, removal of frying debris through frequent removal or filtering, addition of fresh frying medium, and regular cleaning of frying equipment.

9.1.3   EDIBLE FILMS, BATTERS, AND BREADINGS

The oil uptake is conducted through the surface of the food, so the shape and surface characteristics of the food will influence the oil uptake. For instance, French fries can be sliced in larger chunks or cut like cylinders, or surface roughness can be reduced by control of the quality of the slicing blades. Lamberg, Hallstorm, and Olsson (1990) studied these types of techniques.

Since the properties of the surface of the food are most important for fat uptake, the application of a coating is a promising route. This coating can be thin and "invisible" (Gennadios, Hanna, and Kurth, 1997) or can be thick like a batter or breading. The mechanism of action is not clarified, although sometimes the functionality is ascribed to a specific property (Figure 9.2). Often-mentioned properties of coatings in relation to fat uptake are low moisture content, low moisture permeability, thermogelling, and cross-linking. All properties aim to reduce moisture loss and/or modification of the surface structure formed upon frying (Mellema, 2003).

If the outer layer of the product needs to have low moisture content, we can apply low-moisture level coating. Hydrophilic biopolymers can be used in a coating to reduce water loss, so the oil uptake can be reduced also (Pinthus, Weinberg, and Saguy, 1993). Most commercial biopolymer coatings are polysaccharide coatings. For instance, corn zein (Herald et al., 1996) and gellan gum (Williams and Mittal, 1999b).

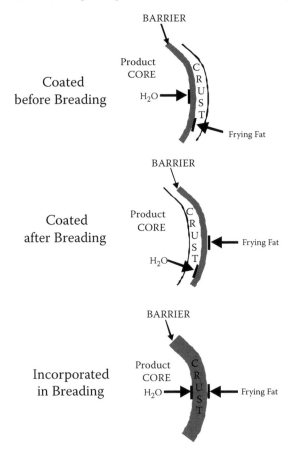

FIGURE 9.2 Projected migration responses of moisture and frying fat to different methods of application of barriers to the product.

Soy protein film coating had also been used for reduction in oil uptake in donuts. Rayner et al. (2000) found that soy protein isolate solutions (10% SPI) with 0.05% gellan gum, as a plasticizer, when cooled after being held at 80°C for 20 min provided suitable films. There was a significant fat reduction (55.12 ± 6.03% on a dry basis) between fried uncoated and coated discs of doughnut mix.

Some coatings were found to be more firm (Pinthus, Weinberg, and Saguy, 1992), so they could reduce evaporation. Often the increase in firmness is brought about by thermogelling action or cross-linking. The resulting high gel strength leads to less evaporation damage and hence to a lower water diffusion (Pinthus et al., 1992). The formation of a large amount of narrow punctures leads to oil uptake because of the high capillary pressures (Moreira et al., 1997). Thermogelling or cross-linking could lead to a stronger, but more brittle coating, which promotes the formation of a small amount of wide punctures with low capillary pressures (Mellema, 2003).

Thermogelling films can be made of MC or HPMC (Meyers, 1994). Garcia et al. (2001) studied the effectiveness of applying an edible coating made of a cellulose derivative. Coatings were applied to French fries by immersion in aqueous suspensions of Methocel A4M. It is important to note in this case that no increase in sogginess was observed by a sensory panel. With most heat-resistant (e.g., cross-linked) water barrier films, evaporation may be too much limited, which will lead to a more soggy fried food. Balasubramaniam, Chinnan, Mallikarjunan, and Phillips (1997) evaluated the effectiveness of hydroxypropyl methylcellulose edible film coating on chicken balls for moisture retention and fat reduction during deep-fat frying. Two sizes (35 and 47 mm) of uncoated and coated chicken balls were fried in a peanut oil at 175°C for five different frying periods ranging from 30 to 180 s for small and 30 to 300 s for large balls. The moisture and fat contents of the surface layer and core were determined. They drew a conclusion of moisture retention up to 16.4% and fat reduction up to 33.7%.

Williams and Mittal (1999b) studied three edible films, gellan gum, MC, and HPMC, and found all these films could reduce the fat absorption into the food by 50 to 91%; however, MC reduced the fat more than the other two films. Gellan gum increased the water loss by about 100%, while HPMC and MC decreased the water loss by 30%. The fat and water diffusivities of the gellan gum were increased with the increase in film thickness.

Batters are defined as "liquid mixtures composed of water, flour, starch, and seasonings, into which food products are dipped prior to cooking," and breadings as "dry mixtures of flour, starch, and seasonings, of coarse composition, and applied to the moistened or battered food products prior to cooking" (Suderman, 1983). Batters can be considered thick coatings. The same principles apply to biopolymer coatings. The main difference is that batters may be more easily applied by consumers. In addition, they have less of the puncturing problems associated with thin coatings (Mellema, 2003).

Commercially, battered or breaded foods are fully or partially cooked by deep-fat frying or oven heating prior to being frozen (Loewe, 1993). Makinson, Greenfield, Wong, and Wills (1987) studied fat uptake during deep-fat frying of coated and uncoated foods and showed fried fish with batters markedly retarded fat movement into the fish. Shih and Daigle (1999) reported that oil uptake of RFBB was less than that of WFBB. Mukprasirt, Herald, Boyle, and Rausch (2000, 2001) found an RFBB for chicken drumstick coating had lower oil absorption than WFBB. The batter-breading coating acts as a barrier to moisture loss and a carrier of spices and other ingredients for flavor enhancement in fried products. Meyers (1990) discussed the functionality of HPMC and MC as barriers in the breading and batters. Muskat and Kerr (2002) reported characteristics of coatings formed from breading of different particle sizes.

Mallikarjunan, Chinnan, Balasubramaniam, and Phillips (1997) evaluated the role of edible coatings to reduce oil uptake in starchy foods like fried potatoes. They used mashed potato balls as the model food system and used corn zein, HPMC, and MC as coatings. Compared to the control, a reduction of 15, 22, and 31% in moisture loss from the products was observed for samples coated with corn zein, HPMC, and MC, respectively. Similarly, a reduction of 59, 61, and 84% in fat uptake in the product crust was observed for samples coated with corn zein, HPMC, and MC, respectively. The difference in performance by the different type of coatings was attributed to the type of coating. Corn zein is a protein-based film whereas the other two are cellulose-based hydrocolloids. The hydrocolloids formed the protective layer due to thermal gelation above 60°C.

The effectiveness of edible coatings formulated from HPMC, MC, corn zein, and amylose in restricting oil absorption during deep-fat frying of akara, a traditional West African food made from whipped cowpea paste, was evaluated for two methods of coating application (spraying and dipping) (Huse, Hung, and McWatters, 2006). Portioned balls of akara paste were partially fried in peanut oil for 100 s at 193°C before applying the coatings. Samples were finish-fried at 166°C to an internal temperature of 70°C.

Although the amylose spray-coated akara was the only treatment resulting in significant reductions in core oil absorption, all treatments absorbed significantly less crust and total oil than the control on a percentage basis. Compared to the control, a 49% reduction in total fat content occurred in dipped MC-coated samples (Table 9.1). While there was no difference between the core moisture retention of treatments, crust and total moisture contents of all treatments were significantly higher than the control samples. Although excess moisture retention of akara may not be desirable because it results in a soggy product, its effects on sensory qualities need to be determined.

As akara cooks, the core temperature rises, and moisture is converted to steam, which is released by the product. The escaping steam forms small capillaries. Oil, in turn, is absorbed into these voids. Therefore, increases in oil content correlate with decreases in moisture content. The decrease in oil absorption is an indirect result of the barrier to moisture removal by the edible coatings. The increase in barrier properties for dipped MC coatings may be due to the increase in film thickness for dipped samples.

MC and HPMC edible film materials were applied to marinated whole chicken strips either prior to breading, after breading, or were incorporated in the breading (Holownia, Chinnan, Erickson, and Mallikarjunan, 2000). Subsequently, the influence of an edible film's location relative to the crust on the fried food product quality and the quality of the frying oil were determined (Table 9.2).

TABLE 9.1

Effect of Edible Coating on Moisture and Fat Contents of Akara Samples

Treatment[1]	Fat Content (%)			Moisture Content (%)		
	Core	Crust	Total	Core	Crust	Total
Control	3.91 ab	54.9 a	31.0 a	57.6 a	31.4 c	44.1 d
Spraying						
HPMC	3.15 abc	34.5 cd	19.6 bc	56.7 a	37.7 b	46.5 c
MC	3.09 abc	41.8 b	21.3 b	57.0 a	34.6 bc	46.4 c
CZ	2.76 bc	35.3 cd	18.6 cd	56.8 a	35.7 bc	46.5 c
Amylose	2.37 c	30.2 e	15.7 f	57.0 a	35.6 bc	46.7 c
Dipping						
HPMC	4.00 ab	34.4 cd	16.5 def	56.8 a	39.6 ab	49.7 b
MC	4.42 a	33.5 d	15.9 ef	57.0 a	44.4 a	52.0 a
CZ	3.02 abc	37.0 c	18.0 cde	57.4 a	35.9 bc	47.9 bc
Amylose	2.76 bc	33.6 d	17.5 cdef	57.2 a	39.7 ab	48.6 bc

Source: Huse, H.L., Hung, Y.-C., and McWatters, K.H. 2006. Physical and sensory characteristics of fried cowpea (Vigna unguiculata L. Walp) paste formulated with soy flour and edible coatings. *J. Food Qual.* 29(4): 419–430.

Note: Mean values in a column not followed by the same letter are significantly different ($\alpha = 0.05$).

[1] HPMC: Hydroxypropyl methyl cellulose, MC: Methyl cellulose, CZ: Corn zein.

Fat and moisture contents of the treated and control core products ranged from 3.4 to 4.5% and 70.0 to 71.3%, respectively (Holownia et al., 2000). No significant effects in fat and moisture content of the core treated products were observed compared to control samples, where no edible coating had been applied. These results dispute those by Mallikarjunan et al. (1997) and Balasubramaniam, Chinnan, and Mallikarjunan (1995, 1997) where significant differences in fat and moisture content were found in fried product preceded by treatment with edible films. It was postulated by the authors that the larger mass of the product used in their studies may have overshadowed any fluxes of fat and moisture that occurred during frying (Holownia et al., 2000).

However, noticeable effects were observed in fat and moisture contents of the product crust following frying of marinated chicken strips to which edible films had been applied. With three of the four coating materials, crust moisture contents decreased if the film was coated prior to breading, whereas when the coating material was incorporated in the breading, moisture contents were not different from those crusts lacking the coating materials. These responses agree with those projected for edible film materials acting as a barrier to water. An inconsistent response was observed when edible coatings were applied after the breading with HPMC E15 coated crusts being slightly lower in moisture, MC A15 coated crusts being slightly higher, and no difference in HPMC E4M or MC A4M coated crusts compared to control crusts. During frying, it was observed that edible coatings applied after the breading tended to flake off, possibly due to reduced binding of the edible film to the breading material as opposed to the chicken surface. This tendency to flake off could be responsible for the inconsistent effects to this mode of application.

TABLE 9.2
Changes in Crust and Frying Oil Properties following Frying of 30 Batches of Marinated Chicken Strips[1]

| Coating | Coating Method | Crust | | | Frying Oil[2] | | | |
		Moisture (%)	Fat (% dry wt)	ΔFFA (% oleic acid)	ΔLovibond Red Value	ΔFood Oil Sensor	ΔViscosity (mPa)
Control[3]	—	28.3 ab	37.1 b	0.08 a	0.83 b	−0.30 b	−0.06 c
HPMC E4M	Before breading	26.7 a	38.7 b	0.04 c	0.63 c	−0.20 c	0 b
MC A4M	Before breading	30.1 b	39.2 b	0.05 b	0.97 a	−0.95 a	0.03 a
HPMC E15	In breading	34.3 c	35.3 a	0.04 c	0.95 a	−0.35 b	0.02 a
MC A15	In breading	33.8 b	34.7 a	0.04 c	0.82 b	−0.05 d	0 b

[1] Values followed by a different letter are significantly different at $p \leq 0.05$.
[2] Change in values from before frying to after frying.
[3] Control – No edible coating applied.

Source: Holownia, K.I., Chinnan, M.S., Erickson, M.C., and Mallikarjunan, P. 2000. Quality evaluation of edible film-coated chicken strips and frying oil. *J. Food Sci.* 65(6): 1087–1090.

The edible films selected for the study by Holownia et al. (2000) were chosen for their ability to inhibit fat migration into the product (Balasubramaniam et al., 1995, 1997). Therefore, in products where edible coatings were applied prior to the breading, it was anticipated that fat in the frying medium would penetrate the crust and in fact, with three of the four edible film materials, crusts had higher fat contents than the control. In contrast, it was anticipated that for those products where edible films would have been incorporated in the breading material, the fat should have been obstructed from entering the crust. Such a process appeared to be the case with the exception of the crust where MC A4M had been evaluated. Incorporation of the edible film material into the breading materials removes an extra step in preparation and, consequently, could be easily adapted to current industrial and institutional preparation procedures. Films applied to chicken strips prior to the breading had fried crusts with higher fat and lower moisture levels. Decreased degradation of frying oils was recorded when product coated with HPMC (food grade E4M) films had been fried compared to non-coated products. It is postulated that these edible films hindered the migration of moisture and acetic acid into the frying oil and this activity was responsible for reduced FFA generation in those oils used to fry the coated products.

9.1.4 MODIFICATION OF FRYING MEDIUM

The frying medium is usually a triglyceride-type oil. The use of triglycerides with polyunsaturated fatty acids is desired from a health point of view. Even though it may influence some quality factors of the fried food like texture and appearance (Brinkmann, 2000), it is widely accepted that the balance of fatty acids does not significantly influence fat uptake. The fatty acid composition of the fat taken up by a fried food does not differ from the fatty acid composition of the frying medium. For high-fat foods, there is a fast exchange of fats with the frying oil during frying. In addition, the use of hydrogenated fat does not significantly affect oil uptake (Mellema, 2003).

A minor positive correlation between oil uptake and oxidative degradation of the frying fat has been identified (Berry, Sehgal, and Kalra, 1999; Dobarganes, Marquez-Ruiz, and Velasco, 2000; Tseng, Moreira, and Sun, 1996). The reason for this is probably a combination of decreased oil/air surface tension and increased oil viscosity (Tseng et al., 1996).

Pinthus and Saguy (1994) reported significant effects of pure PGPR, Tween80, and Span80 as frying mediums on interfacial tension to reduce the wetting of the oil and medium uptake. Assuming well-wetting conditions, the surfactants will give an increased fat uptake and this is proved by the observation of a positive relation between oil degradation and fat uptake (Berry et al., 1999; Dobarganes et al., 2000).

A high oil viscosity, or a steeply increasing oil viscosity upon cooling (especially in the presence of hard fats), will decrease oil uptake because oil flow is hampered especially in the small pores. However, the same will lead to less easy drainage or shaking of the oil from the food after it is taken out of the frying oil. In principle, the total amount of oil adhering to the food determines the maximum amount of oil that can enter the pores (Mellema, 2003). Saguy, Shani, Weinberg, and Garti (1996) studied the oil uptake in different viscosities of oil (jojoba oil and cotton oil) and obtained the same conclusion that in higher viscosity oil (jojaba oil), oil uptake of a deep-fried restructured potato product was significantly higher than in lower viscosity oil (cotton oil).

Until now, no food-grade techniques are available to alter the wetting/surface tension characteristics sufficiently to obtain a significant reduction in fat uptake over the whole timescale. Poly dimethyl siloxanes (PDMS) are used to reduce degradation and it is widely applied in frying oils. It is accepted that it produces a minor reduction in fat uptake.

9.1.5   OTHER METHODS TO REDUCE OIL UPTAKE

Besides the factors mentioned previously that affect the absorption of oil, there are still many factors that can influence oil uptake, such as food particle size, moisture content, and oil temperature. Krokida, Oreopoulou, and Maroulis (2000) studied the relationship between moisture loss and oil absorption during deep-fat frying of French fries. The results showed that oil temperature and thickness of potato strips have a significant effect on oil uptake and moisture loss of French fries. Moreira and Barrufet (1996) reported that oil content of tortilla chips was significantly affected by initial moisture content and particle size distribution. Higher initial moisture content and smaller particle size resulted in higher final oil content. The pore size distribution and mass of air developed during frying results in higher oil content due to high capillary pressure within the pores during cooling.

9.2   ENHANCING CRISPNESS IN BREADED FRIED PRODUCTS

Edible coatings have been researched as a possible means of reducing oil uptake and limiting moisture transfer during frying, thus creating products that are moister and lower in fat. However, limited research has been done to study the effect of edible coatings on crispness. The ability of the edible coatings to limit moisture transfer may enable fried food products to maintain their crispness. To support this theory and to evaluate the effect of edible coatings on product crispness, Ballard and Mallikarjunan (2006) evaluated the crispness in pressure-fried poultry products stored under a food warmer for extended periods. The development of a pressure frying system in which the size of the fry load is not the limiting factor in pressure generation using an inert gas like nitrogen (Innawong, Mallikarjunan, Marcy, and Cundiff, 2006) also may affect the product quality in terms of crispness. Ballard and Mallikarjunan (2006) evaluated the effect of steam or nitrogen on product crispness. Edible coatings such as MC and whey protein isolates (WPI) were used in this study and they were incorporated either into the breading or into the batter. An industrial breading and battering system was used to prepare the samples. Crispness was measured in terms of ultrasonic velocity based on the study by Antonova, Mallikarjunan, and Duncan (2003).

Crispness decreased with increased holding time under the heat lamp, indicating a steady decrease in product crispness (Table 9.3). Pressure source (either steam or nitrogen) in a modified commercial pressure fryer had a significant effect on product

TABLE 9.3
Mean Ultrasonic Velocity as Affected by Pressure Source and Coating Type

Pressure Source[1]	Coating	Ultrasonic Velocity (m/s)
Steam	Uncoated	342.64[ab]
Steam	WPI	528.00[c]
Steam	MC	183.69[a]
Nitrogen	Uncoated	530.90[bc]
Nitrogen	WPI	385.63[b]
Nitrogen	MC	565.97[c]

[1] All samples fried at 163 kPa.

[abc] Means within a column with unlike superscripts are significantly different ($p < 0.05$).

Source: Ballard, T.S. and Mallikarjunan, P. 2006. The effect of edible coatings and pressure frying using nitrogen gas on the quality of breaded fried chicken nuggets. *J. Food Sci.* 71: S259–S264.

crispness. The mean values for crispness in terms of ultrasonic velocity were 485.93 and 353.18 m/s for products fried with nitrogen gas and steam, respectively. These results suggest that samples fried using nitrogen as the pressurizing medium were crispier than samples fried using steam. In addition, the interaction of pressure source and coating type had a significant effect on sample crispness. The mean ultrasonic velocity for the control samples fried using nitrogen gas was 530.9 m/s. Once the MC was added to the nuggets, the mean ultrasonic velocity increased from 530.9 to 566 m/s. The exact opposite occurred when WPI was added to the samples. The addition of WPI decreased the mean ultrasonic velocity from 530.9 to 385.6 m/s. MC-coated samples were significantly crispier than WPI-coated samples. An opposite effect was seen upon evaluating the samples fried using steam as the pressure source. The mean ultrasonic velocity for the MC-coated samples and the control samples decreased when steam was used as the pressure source. In the case of the WPI-coated products fried using steam, the mean ultrasonic velocity was higher than that of MC-coated samples fried using steam. WPI appeared to be more effective at enhancing product crispness when steam was used as opposed to nitrogen gas. These results may be due to some unknown interaction between WPI and nitrogen, which results in a less crisp product when frying using nitrogen. Perhaps the pressure sources had different effects on the film-forming properties of WPI.

Coating application as a main effect had a significant effect on sample crispness. The mean ultrasonic velocity values for the control, samples in which the edible coating was added into the pre-dust, and samples in which the coating was added to the batter were 436.8, 472.3, and 357.2 m/s, respectively. The control samples were not significantly different from the samples in which the edible coating was incorporated into the pre-dust and the batter. However, the samples in which the edible coating was added to the pre-dust were significantly different from the samples in which the coating was added to the batter. Incorporating the coating into the batter decreased the crispness of the samples. Coating application did not have a significant effect on peak load and energy to peak load.

Pressure source as a main effect had a significant effect on sample texture. The mean values for energy to peak load were 2643 and 2556 N × mm for samples fried with steam and nitrogen gas, respectively (Table 9.4). Product fried with nitrogen gas required less energy to reach the peak load. These results are similar to those found by Innawong (2001) in which it was found that frying chicken nugget samples using nitrogen gas resulted in significantly lower energy to peak force and total energy to failure than did products fried with steam. Coating type as a main effect had a significant effect on the energy to peak load. As expected, the energy to peak load increased with increased holding time under the heat lamp due to the moisture loss occurring upon being held under the heat lamp (Figure 9.3). As the product remained under the heat lamp, it became tough and therefore required more energy to reach the peak load. The mean energy to peak load values were 2742, 2457, and 2601 N × mm for the uncoated samples, WPI-coated samples, and MC-coated samples, respectively. The control samples were significantly different from both the WPI- and MC-coated samples, requiring the most energy to reach the peak load. The WPI-coated samples had a lower mean energy to peak load value.

In essence, the type of ingredients in the batter and breading can influence the oil uptake in fried foods. The food industry is trying to explore other protein-based ingredients like wheat and fish protein mix in a variety of fried foods in order to reduce the oil uptake. Many times, the reduction in oil uptake at the expense of reducing moisture migration could result in a soggy or tough product during storage under the heat lamp. Therefore, the selection of breading and batter ingredients becomes crucial and requires extensive research.

TABLE 9.4
Mean Peak Load as Affected by Pressure Source and Coating Type

Pressure Source[1]	Coating	Peak Load (N)
Steam	Uncoated	477[bc]
Steam	WPI	515[d]
Steam	MC	461[b]
Nitrogen	Uncoated	516[d]
Nitrogen	WPI	419[a]
Nitrogen	MC	489[c]

[1] All samples fried at 163 kPa.

[abcd] Means within a column with unlike superscripts are significantly different ($p < 0.05$).

Source: Ballard, T.S. and Mallikarjunan, P. 2006. The effect of edible coatings and pressure frying using nitrogen gas on the quality of breaded fried chicken nuggets. *J. Food Sci.* 71: S259–S264.

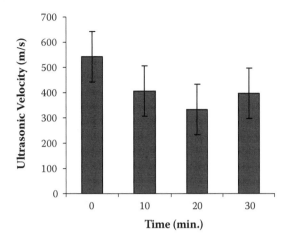

FIGURE 9.3 Mean ultrasonic velocity and mean peak load of samples held under a heat lamp (60°C) at different time intervals. (*Source*: Ballard, T.S. and Mallikarjunan, P. 2006. The effect of edible coatings and pressure frying using nitrogen gas on the quality of breaded fried chicken nuggets. *J. Food Sci.* 71: S259–S264.)

9.3 REFERENCES

Aguilera, J.M. and Gloria, H. 1997. Determination of oil in fried potato products by differential scanning calorimetry. *J. Agric. Food Chem.* 45: 781–785.

Antonova, I., Mallikarjunan, P., and Duncan, S.E. (2003) Correlating objective measurements of crispness in breaded fried chicken nuggets with sensory crispness. *J. Food Sci.* 68(4): 1308–1315.

Balasubramaniam, V.M., Chinnan, M.S., and Mallikarjunan, P. 1995. Deep-fat frying of edible film coated products: Experimentation and modeling. *Food Process. Automation IV.*

Balasubramaniam, V.M., Chinnan, M.S., Mallikarjunan, P., and Phillips, R.D. 1997. The effect of edible film on oil uptake and moisture retention of a deep-fat fried poultry product. *J. Food Proc. Eng.* 20(1): 17–29.

Ballard, T.S. and Mallikarjunan, P. 2006. The effect of edible coatings and pressure frying using nitrogen gas on the quality of breaded fried chicken nuggets. *J. Food Sci.* 71: S259–S264.

Berry, S.K., Sehgal, R.C., and Kalra, C.L. 1999. Comparative oil uptake by potato chips during frying under different conditions. *J. Food Sci. Technol.* 36: 519–521.

Brinkmann, B. 2000. Quality criteria of industrial frying oils and fats. *Eur. J. Lipid Sci. Technol.* 102: 539–541.

Debnath, S., Bhat, K.K., and Rastogi, N.K. 2003. Effect of pre-drying on kinetics of moisture loss and oil uptake during deep fat frying of chickpea flour-based food. *Lebensm.-Wiss.U.-Technol.* 36: 91–98.

Dobarganes, C., Marquez-Ruiz, G., and Velasco, J. 2000. Interactions between fat and food during deep-frying. *Eur. J. Lipid Sci. Technol.* 102: 521–528.

Garcia, M.A., Ferrero, C., Bertola, N., Martino, M., and Zaritzky, N. 2001. Effectiveness of edible coatings from cellulose derivatives to reduce fat absorption in deep fat frying. Abstract #73C-33, IFT Annual Meeting, New Orleans, LA.

Gennadios, A., Hanna, M.A., and Kurth, L.B. 1997. Application of edible coatings on meats, poultry and seafoods: A review. *Lebensm. Wiss. Technol.* 30: 337–350.

Herald, T.J., K.A. Hachmeister, S. Huang, and J.R. Bowers. 1996. Corn zein packaging materials for cooked turkey. *J. Food Sci.* 61: 415.

Holownia, K.I., Chinnan, M.S., Erickson, M.C., and Mallikarjunan, P. 2000. Quality evaluation of edible film-coated chicken strips and frying oil. *J. Food Sci.* 65(6): 1087–1090.

Huse, H.L., Hung, Y.-C., and McWatters, K.H. 2006. Physical and sensory characteristics of fried cowpea (*Vigna unguiculata L. Walp*) paste formulated with soy flour and edible coatings. *J. Food Qual.* 29(4): 419–430.

Iles, B.C. and Elson, C.R. (1972). Crispness of cereal. In: *Advances in cereal science and technology*. Pomeranz, Y. (Ed.). St. Paul, MN: American Association of Cereal Chemists, pp. 1–19.

Innawong, B. 2001. Improving fried product and frying oil quality using nitrogen gas in a pressure frying system. Doctoral dissertation. Virginia Polytechnic Institute and State University, Blacksburg, Virginia.

Innawong, B., Mallikarjunan, P., Marcy, J., and Cundiff, J. 2006. Pressure conditions and quality of chicken nuggets fried under gaseous nitrogen atmosphere. *J. Food Proc. Preserv.* 30(2): 231–245.

Katz, E.E. and Labuza, T.P. 1981. Effect of water activity on the sensory crispness and mechanical deformation of snack food products. *J. Food Sci.* 46: 403–409.

Kolesch, C. 1994. Edible water vapor barriers: Properties and promise. *Trends Food Sci. Technol.* 5: 76–81.

Krochta, J.M. and Mulder-Johnston, C.D. 1997. Edible and biodegradable polymer films: Challenges and opportunities. *Food Technol.* 51(2): 61–74.

Krokida, M.K., Oreopoulou, V., and Maroulis, Z.B. 2000. Water loss and oil uptake as a function of frying time. *J. Food Eng.* 44: 39–46.

Lamberg, I., Hallstorm, B., and Olsson, H. 1990. Fat uptake in a potato drying frying process. *Lebensmittel-Wissenschaft Technologie*, 23: 295–300.

Loewe, R. 1993. Role of ingredients in batter systems. *Cereal Foods World* 38(9): 673–677.

Makinson, J.H., Greenfield, H., Wong, M.L., and Wills, R.B.H. 1987. Fat uptake during deep-fat frying of coated and uncoated foods. *J. Food Comp. Anal.* 1: 93–101.

Mallikarjunan, P., Chinnan, M.S., Balasubramaniam, V.M., and Phillips, R.D. 1997. Edible coatings for deep-fat frying of starchy products. *Lebensmittel-Wissenschaft und-Technologie*, 30(7): 709–714.

Maskat, M.Y. and Kerr, W.L. 2002. Coating characteristics of fried chicken breasts prepared with different particle size breading. *J. Food Proc. Preserv.* 26: 27–38.

Mehta, U. and Swinburn, B. 2001. A review of factors affecting fat absorption in hot chips. *Crit. Rev. Food Sci. Nutr.* 41: 133–154.

Mellema, M. 2003. Mechanism and reduction of fat uptake in deep-fat fried foods. *Trends Food Sci. Tech.* 14: 364–373.

Meyers, M.A. 1990. Functionality of hydrocolloids in batter coating system. In: *Batters and breadings in food processing*. Kulp, K. and Loewe, R. (Eds.). St. Paul, MN: American Association of Cereal Chemists. pp. 117–142.

Moreira, R.G. and Barrufet, M.A. 1996. Spatial distribution of oil after deep-fat frying of tortilla chips from a stochastic model. *J. Food Eng.* 27(3): 279–290.

Moreira, R.G. and Barrufet, M.A. 1998. A new approach to describe oil absorption in fried foods: A simulation study. *J. Food. Eng.* 35: 1–22.

Moreira R.G., Sun, X., Chen, Y. 1997. Factors affecting oil uptake in tortilla chips in deep-fat frying. *J. Food Eng.* 31(4): 485–498.

Mukprasirt, A., Herald, T.J., Boyle, D.L., and Rausch, K.D. 2000. Adheion of rice flour-based batter to chicken drumsticks evaluated by laser scanning confocal microscopy and texture analysis. *Poultry Sci.* 79: 1356–1363.

Mukprasirt, A., Herald, T.J., Boyle, D.L., and Boyle, E.A. 2001. Physicochemical and microbiological properties of selected rice flour-based batters for fried chicken drumsticks. *Poultry Sci.* 80: 988–996.

Pinthus, E.J. and Saguy, I. 1994. Initial interfacial tension and oil uptake by deep-fat fried foods. *J. Food Sci.* 59(4): 804–807.

Pinthus, E.J., Weinberg, P., and Saguy, I.S. 1992. Gel-strength in restructured potato products affects oil uptake during deep-fat frying. *J. Food Sci.* 57: 1359–1360.

Pinthus, E.J., Weinberg, P., and Saguy, I.S. 1993. Criterion for oil uptake during deep-fat frying. *J. Food Sci.* 58(1): 204–211.

Pinthus, E.J., Weinberg, P., and Saguy, I.S. 1995. Oil uptake in deep fat frying as affected by porosity. *J. Food Sci.* 60(4): 767–769.

Rayner, M., Ciolfi, V., Maves, B., Stedman, P., and Mittal, G.S. 2000. Development and application of soy-protein films to reduce fat intake in deep-fried foods. *J. Sci. Food Agric.* 80: 777–782.

Rimac-Brncic, S., Lelas, V., Rade, D., and Simundic, B. 2004. Decreasing of oil absorption in potato strips during deep fat frying. *J. Food Eng.* 64(2): 237–241.

Rohm, H. (1990). Consumer awareness of food texture in Austria. *J. Texture Studies* 21: 363–373.

Roudaut, G., Dacremont, C., Valles Pamies, B., Colas, B., and Le Meste, M. 2002. Crispness: A critical review on sensory and material science approaches. *Trends Food Sci. Technol.* 13: 217–227.

Saguy, I.S. and Pinthus, E.J. 1995. Oil uptake during deep-fat frying: Factors and mechanism. *Food Technol.* 49(4): 142–145, 152.

Saguy, I.S., Shani, A., Weinberg, P., and Garti, N. 1996. Utilization of jojoba oil for deep-fat frying of foods. *Lebensm-Wiss.U.Techonol* 29: 573–577.

Shih, F.F. and Daigle, K.W. 1999. Oil uptake properties of fried batters from rice flour. *J. Agric. Food Chem.* 47: 1611–1615.

Southern, C.R., Chen, X.D., Farid, M.M., Howard, B., and Eyres, L. 2000. Determining international oil uptake and water content of fried thin potato crisps. *Food Bioproducts Proc.* 78: 119–125.

Suderman, D.R. 1983. Use of batters and breadings on food products: A review. In: *Batter and breadings.* Suderman, D.R. and Cunningham, F.E. (Eds.). Westport, CT: AVI Publishing Co., Inc.

Tseng, Y.C., Moreira, R., and Sun, X. 1996. Total frying-use time effects on soybean-oil deterioration and on tortilla chip quality. *Int. J. Food Sci. Technol.* 31(3): 287–294.

Ufheil, G. and Escher, F. 1996. Dynamics of oil uptake during deep-fat frying of potato slices, *Lebensm.-Wiss. u-Technol.* 52: 640–644.

Williams, R. and Mittal, G.S. 1999a. Low-fat fried foods with edible coatings: Modeling and simulation. *J. Food Sci.* 64: 317–322.

Williams, R. and Mittal, G.S. 1999b. Water and fat transfer properties of polysaccharide films on fried pastry mix. *Lebensm.Wiss.u.Technol.* 32: 440–445.

Abbreviations

A

ANN: artificial neural network

B

B.C.: boundary conditions
BU: Brabender units

C

CIP: cleaning in place
CMC: carboxyl methylcellulose
CZ: corn zein

D

deep-fat fried
DSC: differential scanning calorimetric

F

French fry
FFA: free fatty acids
FOM: food oil monitor
FTIR: Fourier transformation infrared spectroscopy

G

GC: gas chromatography
GC/MS: gas chromatography/mass spectrometry
GLC: gas liquid chromatography

H

HPLC: high performance liquid chromatography
HPMC: hydroxymethylcellulose
HPSEC: high-performance exclusion chromatography

I

I.C.: initial conditions
IFT: interfacial tension
IR: infrared

M

MC: methylcellulose
MC: moisture content
MCP: monocalcium phosphate

N

NIR: near infrared

O

OC: oil content

P

PC: powdered cellulose
PDMS: poly dimethyl siloxanes
PGPR: polyglycerol polyricinoleate
PSD: pore size distribution
PT: pasting temperature
PV: peak viscosity

R

RFBB: rice flour-based batter

S

SALP: sodium aluminum phosphate
SAPP: sodium acid pyrophosphate
SAW: surface acoustic wave
SEM: scanning electronic microscopy
SFE: supercritical fluid extraction

T

TPC: total polar component
TOF: time-of-flight
TPM: total polar material

W

WFBB: wheat flour-based batter
WPC: whey protein concentrate
WPI: whey protein isolates

Index

Printed and bound by CPI Group (UK) Ltd, Croydon, CR0 4YY

21/10/2024

01777084-0005